Drip Irrigation
For Every Landscape and All Climates

by
Robert Kourik

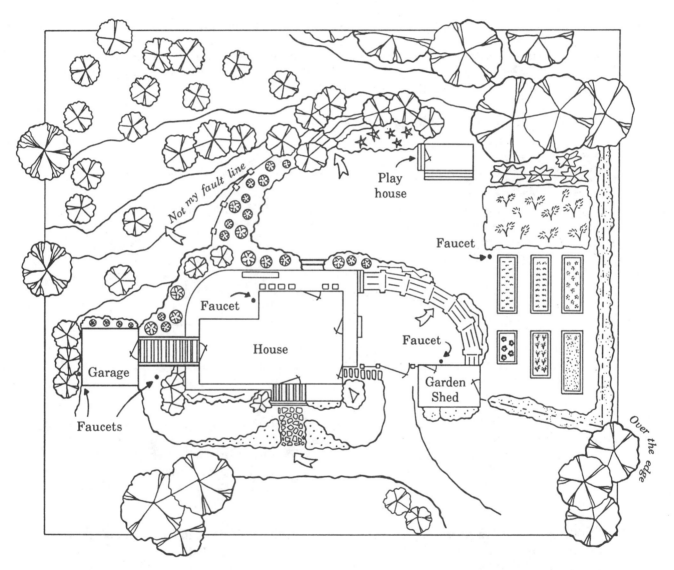

Helping your garden flourish, while conserving water!

Illustrations, design and layout by Heidi Schmidt

Metamorphic Press

Front Cover: An edible landscape designed and photographed by the author for the Preston Vineyards and Winery, Healdsburg, Ca. Except for the small lawn "piazza," the entire landscape is irrigated with in-line drip tubing. This edible ornamental landscape was planned and planted by zones of differing water use. With an automatic controller, each zone receives a different length of irrigation, a different amount of water. As this photograph shows, drip irrigation offers the benefit of being virtually invisible.

Front and back cover: All photographs copyrighted by the author.

Kourik, Robert.
    Drip irrigation for every landscape and all climates: helping your garden flourish while conserving water / Robert Kourik
    p. cm.
    Includes bibliographical references, suppliers and index.
    ISBN 0-9615848-2-3 (pbk.)
    1. Trickle irrigation. 2. Water conservation. 3. Low volume irrigation. 4. Water — plant growth. I. Title.
    631.74 1992   92-228499
                                                              LCCN

Sales to the U.S. book trade, contact:
Independent Publisher's Group, Inc.
814 North Franklin St.
Chicago, IL 60610
(312) 337-0747

Special markets, irrigation stores and companies, mail order sales and the nursery trade, contact:
The Drip Irrigation Book Project
Metamorphic Press
P.O. Box 1841, Santa Rosa, CA 95402
(707) 874-2606

Sales to the book trade in Australia, contact:
Tower Books
Unit 9/19 Rodborough Road
Frenchs Forest
NSW 2086 Australia
(02) 975 5566

Sales to the book trade in Great Britian, contact:
Eco-Logic Books
8 Hunters Moon
Dartington
TOTNES TQ9 6JT England
(0803) 867546

Printed in the United States of America on paper made from 10% post-consumer waste paper. Cruelty-free, not tested on animals. Vegetarian paper. Recyclable. Totally groovy.

10 9 8 7 6 5 4 3 2

Table of Contents

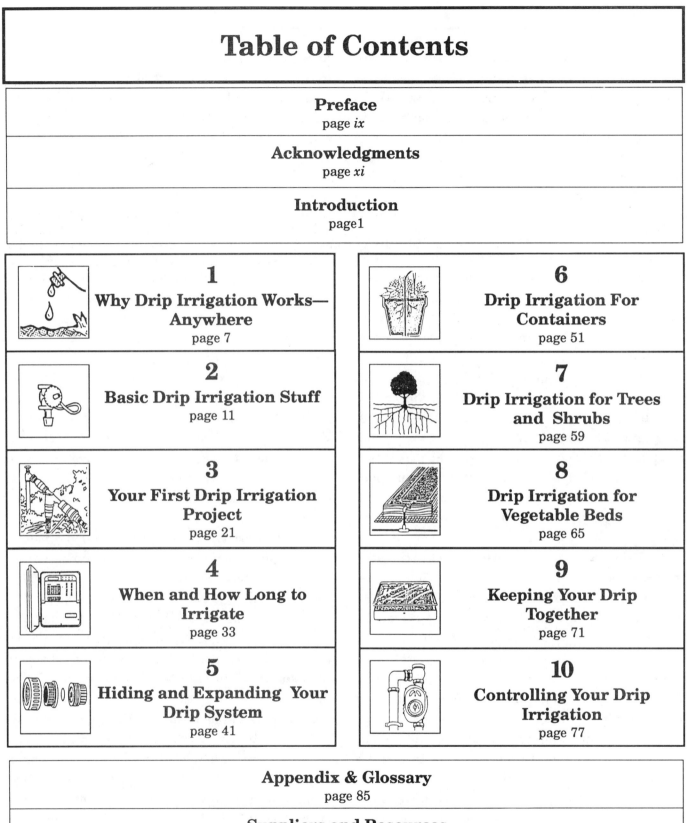

Preface

I wrote this book on drip irrigation for three reasons: because no other book has been published about this water-conserving approach to irrigation, because plants grow *better* with drip irrigation systems than with any other typical watering method and because drip irrigation is a cost- and time-effective tool for watering in *any* climate. (Well, okay—four reasons. The last being to make a little money while doing something good for the environment.)

Many of the pamphlets stuffed into the boxes of those drip irrigation kits found in mega-chain hardware stores read as if they were first written in Arabic, translated into French and then transcribed into English. Even do-it-yourself car-care manuals are easier to read. Seems to me, it's time there is a drip irrigation guidebook which isn't cloaked in overly technical jargon or befuddled by incoherent English. So, while I won't skirt plumbing terminology, this book is meant to guide you carefully and gradually through every piece of a complete drip irrigation system—what it's called (how a plumber would write the part on her parts list) and how to assemble it with the next part. Every part you'll need appears in an illustration. So if you're confused, just point and tell the hardware store clerk that you want "one of these."

Drip irrigation does conserve plenty of water; estimates and studies range from 50% to 70% savings. Conserving precious fresh water supplies has become a national issue. Water conservation is no longer the concern of just a few isolated, drought-stricken communities in the West. While water may not always be scarce everywhere in the country, a supply of fresh, clean water is no longer a given. You needn't be a strident environmentalist to conserve water. Making good use of what water our landscapes and gardens soak up is just being thrifty and wise. The savings are more far-reaching than just the water which dribbles ever so gently out of your drip irrigation system. For water is closely linked to energy. In most cases, virtually every step of the recovery, purification and distribution of water consumes fuel. To conserve water is to preserve fossil fuels.

For the home gardener, the bottom line is the health of the landscape and garden. Here, drip irrigation's value shines far above all other watering methods. It is a less well known fact that drip-irrigated plants usually grow *better* than those sprayed by hand or sprinklers. Trees grow faster and wider. Shrubs and perennial flowers bloom more profoundly. Vegetables and fruit and nut trees yield more abundantly. And all this with less fungal disease, mildews or rusts, due to the drier foliage of the drip-irrigated plant.

This book is not an encyclopedia of drip irrigation hardware and design. I write best about what I do. And boy, do I also know what *not* to do! Over the years, I've made every mistake mentioned in this book—and more. And I've manifested every one of Pop Murphy's Laws of Plumbing (see pages 4 & 5)—some more than once. So, this book focuses on what has worked the best for me and those designs which make my plants the happiest. All this has come about after years of learning by trial and error. If this book saves you from the expense and frustration of even 10% of what I've gone through, then your money has been well spent. If, after reading this book, you feel drip irrigation is not for you, then you've saved a fortune in aspirin alone.

Everybody operates differently. I'm a hands-on kind of guy. When writing about plumbing and drip irrigation, I need to have the parts strewn about my desk to make sure I keep everything straight. And I have a very hard time remembering the "trade" name for many of the parts. I just go to the plumbing store, rummage through the shelves for what I need and grab some stuff. So, if you're like me, don't be intimidated by the drawings and names of the parts in this book. Just take the book to your local plumbing or hardware store and select a bunch of parts to match the drawings. Then keep those parts handy while reading the book and finger-fit the various parts together as they're mentioned in the text. Play with your parts. See how many ways you can fit them together. Get familiar with the parts *before* you're bent over on a hot and dry June afternoon with salty sweat dribbling into your eyes and your sciatica begining to electrify nerve endings you never knew existed. Do your homework before you literally get in the trenches.

No matter what seems to be going wrong, laugh. Or call a plumber. But first check the equity in your house! Remember: What does a plumber charge when one of the pipes you threaded together spouts a major leak, it's three o'clock in the morning and your wife is going into labor? Anything she wants!

But seriously, folks, you'll master drip plumbing, see the benefits grow right before your eyes and proudly award yourself the Degree in Dripology found at the end of this book.

Robert Kourik

Robert Kourik

Occidental, June 1992

Acknowledgments

As with the creation of any book, there are many people who have offered personal and professional support. Most books are produced with the help of a veritable social web of friends, associates and technical authorities—and this book is no exception. Many books are dedicated to a single person or a few important people, but I'd like to dedicate this book to all those listed below and to the greater domain of widespread friendship which makes publishing possible. The following alphabetical listing is meant to be fairly complete. But if there's anyone who feels left out, please accept my apologies and thanks.

Chester Aaron—A good buddy and steadfast friend with whom I can commiserate on the difficulties and injustices in the publishing world—not to mention our shared love of garlic, lots of garlic.

David Baldwin and **Karin Kramer**, The Natural Gardening Company—The folks who trusted me enough to allow me to develop a drip irrigation system for their mail-order catalog. They know my many foibles and continue to be supportive friends and colleagues.

Elvin Bishop—A great blues guitarist and gardener. His r&b has nursed me though many tough publishing troubles. Also for teachin' me how to land a few fish.

Mary Burke—A client and neighbor who has always been very tolerant of my scattered schedule as publishing demands gradually cut into my landscaping work. Her garden was the site of my last drip irrigation installation.

Betsy Clebsch—A wise gardener, dear friend and steadfast advocate of horticulture.

Sandy Farkas—The accomplished artist who designed the front and back covers of the book. Someone who was professional and understanding throughout the production of my "edible" book and gladly returned to work on this book.

Pat Harris—My copy editor, who possesses an amazing amount of attention to detail and applied this (to me) mysterious and valuable skill to the copyediting of the final text. Any mistakes in the text are from my neglect or error, certainly not hers.

David Henry, Certified Irrigation Designer—The drip irrigation consultant who pored over the raw manuscript for technical errors. He has taught me many drip irrigation design guidelines and facts and the important mantra, "Check-valve, filter, then pressure regulator." Any technical errors which slipped into the book are the fault of the author, not David.

Amie Hill—The patient editor who worked on my rough manuscripts as I dropped them off at her house in my willy-nilly, erratic fashion. Not only did she greatly improve my grammar, syntax and sentence structure, she also always left helpful notes explaining my most repetitious linguistic blunders.

Ron and Diane Hudelson—For their generous investment in this project.

Dave Johnson—Who generously allowed me to use some charts from his eye-opening and thoroughly researched 1977 booklet titled "Water Requirements of Drought-Resistant Plants." One of the early pioneers in ET and dry-farming literature research.

Michele Kliphon—The reliable and able bookkeeper who has computerized the shipping and bookkeeping for all my book projects. Who persevered through the era, or error, of the "Too many files open" prompt in the accounts receivable program.

Rosa May and **John Kourik**—My mom and dad have been unwavering in their support during this project. Their financial and emotional support has much to do with this book's even being born. Plus, it's fun when they come out to visit.

Elinor Lindheimer—A pro at the unusual, but essential, art of indexing. She patiently tolerated my in-house management mistakes to produce yet another accurate and highly accessible index. An entry in the Index should say: "Accuracy—see also Lindheimer."

Marshia Loar—The patient and understanding landlord for my home, office and "warehouse." The care of her property sometimes languishes as I become embroiled in final-production obligations. Without her support for "one of the small guys" in publishing, I don't think I could continue to publish.

Kurt Maloney—The landscape representative from Netafim who reviewed the manuscript for technical errors and provided much-needed support.

National Bank of the Redwoods—Thanks to Pat Kilkinney and Jerry Johnson for having faith in my business, in spite of the lack of both traditional hard assets and a conventional business structure. They have helped me form a sound financial foundation. This is a modern financial institution with all the warmth, familiarity and flexibility of an old-fashioned small-town bank.

Linda Parker—Special thanks for some drawings so skillfully provided under pressure at the very end of the project. Also, the most painterly bookkeeper I know.

Myra Portwood—A saint of a client who patiently withstood the vacillations of my career as writing and publishing slowly, and unpredictably, replaced landscape installation. Also, a mutual lover of lavender.

Dave Roberts—For his expertise in computers and his investment in this project.

Bruce Robinson, Desktop Plus—Assisted in the final details of computerized production of the galleys. Cheerfully entered all the finishing touches under the pressure of the final deadline.

John Schmidt—One of the northern California representatives for Netafim who helped make sure this book is accurate.

Dave Smith—One of the people who originally supported this project. His steadfast encouragement has been most helpful over the years. The consummate representation of an ethical and considerate businessman.

Molly Sterling—My steadfast and attentive therapist. With her help, I've been able to avoid repeating some of the errors of my past. Thanks.

Jim Sullivan—Who made sure my atmospheric vacuum breakers weren't on backward. A skilled landscape designer and political commentator who also reviewed the entire text for general goofs.

Summerfield Graphics—For the excellent work on the color separations and final filmwork for the covers.

Jae Treesinger—For her thoughtful conversation during a much-needed retreat and refuge along the great north coast during production of this book.

The Union Bakery, Occidental—Thanks to Shelley, Barbara, Jessica and all the folks at the local coffee shop and bakery—their lattes helped fuel many an early morning writing effort.

Herman Warsh and **Maryanne Mott**—For their generous investment in this project.

Neil Wilkinson, Delta Lithograph Company—The salesman who, along with Pam Haus, carefully guided the book through the printing process. An avid gardener who has listened to all my deadline-induced ravings and still remained friendly and helpful.

Introduction

Somewhere out there, perhaps, is a garden that *wouldn't* benefit from drip irrigation. In this garden, the rains come with absolute regularity, just when the soil is beginning to dry, and never too far apart. For *real* gardens, however, watering between irregular rains or during protracted droughts is a fact of life.

Although drip systems are initially somewhat more complicated than dragging a hose around with an attached oscillating sprinkler, the benefits from drip irrigation, in *any* climate, far outweigh the hassles.

Is Drip for You?

Drip irrigation isn't for everybody or every situation. No technology or tool is appropriate for universal application, and like any gardening tool, drip irrigation has various benefits and certain limitations. Before you decide to design and install your own system, consider the following pros and cons of drip irrigation. In reading through them, you may find that the drawbacks outweigh the advantages for your garden, in which case an extensive drip system probably isn't for you. Most gardeners, however, will find that drip irrigation may be appropriate for a portion, if not all, of their ornamental plantings and at least some of their edible crops.

The Benefits of Drip Irrigation

The advantages of drip irrigation can be summarized as follows:

● **Uses water efficiently.** Sprinklers waste a lot of water as a result of wind-scattered spray, sun-powered evaporation, runoff, the evaporation of accumulated puddles or deep leaching.

● **Provides precise water control.** Every part of a drip irrigation system can be constructed with an exact flow rate. It is very easy to calculate what the total flow of the system amounts to and to match this with the needs of plants. You'll know exactly how much water you're applying and will be able to control the amount down to the ounce.

● **Increases yields.** Drip irrigation can be used for the slow, gradual application of tiny amounts of water on a frequent, or daily, basis. This maintains an ideal soil moisture level, promoting more abundant foliage, greater bloom, and higher yields (by actual comparison) of produce, fruits and nuts than those produced by any other type of irrigation.

● **Provides better control of saline water.** Sprinklers apply water to the foliage; if your water is saline, this can cause leaf burn. Drip irrigation applies water only to the soil, and frequent applications with drip irrigation help to keep the salts in solution so they don't affect the roots adversely. (Any salt crust buildup at the margins of the moist area can be leached away with an occasional deep irrigation.)

● **Improves fertilization.** With a device called a fertilizer injector (or proportioner), you can easily apply dissolved or liquid fertilizers with accuracy and without leaching the fertilizer beyond desired root zones. The liquid fertilizers can be applied with each irrigation or only when required.

● **Encourages fewer weeds in dry-summer climates.** The small moist spot around each emitter, where the water slowly dribbles out, covers only a fraction of the soil's surface. The larger dry areas between emitters remain too dry for weed seeds to sprout. (Note: This benefit is lost in areas with summer rains.)

● **Saves time and labor.** Drip irrigation systems eliminate tedious and inefficient hand watering. Automatic drip systems add the convenience of your not even having to remember to turn valves on and off by hand. (The initial installation of such a system, however, will take more time and effort than all other forms of irrigation except permanent sprinkler systems.)

● **Reduces disease problems.** Without the mist produced by a sprinkler, drip-irrigated plants are less likely to develop water-stimulated diseases such as powdery mildew, leaf spot, anthracnose, shothole fungus, fireblight and scab. Furthermore, careful placement of emitters away from the trunks of trees, shrubs, perennials and vegetables will keep the crown of the root system dry and minimize such root problems as crown rot, root rot, collar rot and armillaria root rot.

• **Provides better water distribution on slopes.** Sprinklers often create wasteful runoff when set to water the upper slopes of hills or berms. Drip emitters can apply the water slowly enough to allow all the moisture to soak into the soil. Some emitters, known as pressure-compensating emitters, are designed to regulate the flow of water so that all emitters in the system put out the same gentle flow, regardless of slope.

• **Promotes better soil structure.** Heavy sprinkler irrigation can produce puddles cause clay particles to stick together and increase soil compaction. Drip-applied water gradually soaks into the ground and maintains a healthy aerobic soil which retains its loamy structure.

• **Conserves energy.** Because of the low pressure requirements of a drip irrigation system the pumping costs, whether from municipal water supplies or from your own well, are lower.

• **Uses low flow rates.** The low-volume application rate of drip emitters permits larger areas to be watered at the same time than is possible with sprinkler systems.

• **Is more economical than permanent sprinkler systems.** While more costly than a hose with an oscillating or hand-held sprinkler, drip irrigation systems usually cost less than fixed sprinkler systems.

The Limitations of Drip Irrigation

Some drawbacks of drip irrigation include the following:

• **Eliminates soothing hand watering.** For some gardeners, the act of standing out in the garden near sunset and watching the moon rise, listening to the mockingbird warble and rhythmically swaying back and forth with a hand-held sprinkler is more valuable than any form of therapy or meditation. For these people, drip irrigation may be counterproductive.

• **Initial costs are high.** A garden hose with a simple oscillating sprinkler will always be cheaper than drip irrigation, but it doesn't offer the same measure of control and water conservation. A well-designed drip system will more than repay the cost of installation in reduced effort, fewer irrigation chores and greater yields.

• **Can clog.** This is really a limitation of older and outdated emitters. Many early models of emitters were more prone to clogging and gave the industry a bad reputation. All this has changed; with adequate filtration, the emitters featured in this book have performed successfully with city water, well water and water with dissolved iron and calcium. Some types, such as the noncompensating in-line emitter tubing, even function reliably with filtered gray water systems.

• **May restrict root development.** Early and out-moded designs for drip systems called for only one or two emitters per plant. This led to very restricted root growth around the few points of moisture and thus to stunted plant growth. One major goal of this book is to show how the emitters should be placed to cover the entire area of natural root growth. With the proper placement of emitters, root growth will be uniform, expansive and healthy.

• **Rodents can eat the tubing.** There are a number of drip irrigation systems which can be buried in the soil, but if your garden has gophers, you'll just be offering them an easier way to drink. Occasionally even mice and wood rats will hear the running water in the drip hose, especially in dry summer climates, and chew through for a swig of *aqueous delecti*.

• **Isn't compatible with green manure and cover crops.** The growth of a green manure crop gets all tangled up with the drip tubing, thus prohibiting the usual tilling under of the plants. Cover crops are usually grazed or mown, both of which would damage a surface drip system. And both buried and surface drip systems would leave too much of the surface dry for adequate germination of cover crop or green manure seeds.

• **Weeding can be difficult.** Unmulched drip irrigation systems will stimulate some weeds around each emitter, and care must be taken not to damage the drip system while weeding. A protective and attractive layer of mulch will greatly reduce, if not eliminate, this problem.

• **Requires greater maintenance.** Very little can go wrong with your typical hose-and-oscillating-sprinkler setup. Drip irrigation requires more routine maintenance to sustain its high level of efficiency, but such maintenance is relatively simple.

• **Doesn't cleanse the foliage.** In arid climates, some plants, such as lettuce and other leafy greens, prefer periodic sprinkling of leaves to wash off accumulated dust, grit and/or air pollution. These plants need to receive an occasional sprinkling or be grown with low-flow sprinklers.

• **Doesn't create humidity.** Many plants, most notably humidity-loving perennials from England, the muggy tropics and northern Europe, like a moist atmosphere. When these plants are grown outside their natural environment, sprinklers and misters are perhaps the better irrigation devices. (For many other plants, a drip system's lack of additional humidity is a bonus, inhibiting fungal and bacterial diseases.)

• **You can't see the system working.** With a well-mulched drip system, the emitters quietly go about their work hidden from view. For some gardeners this is the beauty of the system. For others, not being able to watch the watering is slightly unsettling. In a poorly-designed system, a clogged emitter goes unnoticed until drought stress affects the plant visibly. This is a serious problem only with poor quality emitters and the old-fashioned concept of placing only one or two emitters next to each plant. The design approach in this book protects the plants from any stress as a result of clogging.

Simple, but Not Easy

People have good cause to be leery about drip irrigation. At first, it seems so darn complicated. Yet over and over, the phrase "simple, but not easy" comes to my mind.

I got this description from Warren Schultz, editor-in-chief for *National Gardening* magazine, as we were lamenting the journalistic difficulties of explaining drip irrigation. I like his simple-but-not-easy concept because it capsulizes the fact that the basics of drip irrigation really aren't that difficult to understand or implement—provided you take the time to do a little homework.

There are, however, a number of initial stumbling blocks that prevent many people from even considering messing with this complicated-looking drip business.

Unless it's done with planning, a sense of craftsmanship, and a desire to achieve that elusive look of affordable elegance, a drip irrigation system can be just more garden clutter, and *ugly* to boot, with black plastic spaghetti tangles running from plant to plant.

After years of experimenting, my approach to drip irrigation is to use the least number of different parts for the most efficient system. This book is about how to put in a drip irrigation system that's simple to install, efficient and virtually invisible to the eye.

Gizmophobia

Drip irrigation equipment is pretty weird looking. With gadgets, tubing, dials, buttons, and digital-readout displays, it's enough to send a simple gardener scuttling back to the nineteenth century.

If you're intimidated by any device more complicated than a hand-operated can opener, rest assured that drip irrigation is nowhere near as bad as it seems. Think of all those put-'em-together parts as toys for grown-ups. Harken back to days gone by, the many happy hours spent playing with Lincoln Logs™, Erector Sets™, or Tinker Toys™. If you can put yourself in that same carefree, frame of mind out in your landscape, drip parts may not seem so intimidating.

Fear of Plumbing

Many homeowners are possessed by a phobia more overpowering than the repulsiveness of spiders or the apprehension caused by visiting inlaws—they have a deep-seated fear of plumbing. This syndrome usually begins when young children are allowed to watch their father turn pale upon the receipt of his first bill from a union plumber.

While designing and installing an efficient and effective plumbing system for an entire house is a complicated job and often requires professional skills, the plumbing required for landscaping is well within the means and skills of most gardeners. Anyone can learn the relatively simple craft of garden plumbing. Whatever mistakes you make in plumbing a drip system are easily rectified and virtually harmless, as long as you have correctly installed a backflow preventer (described in detail on pages 11-13 & 46) to protect the purity of your home's drinking water.

The Truth About *%&#$ Plumbing

One important plumbing skill is the well-hurled curse, whether unspoken or thunderously roared. Anyone who proposes to instruct you in trouble-free drip irrigation installation and assures a carefree, routine process is either a liar or a terminal optimist. For they are forgetting the most persistent and, perhaps, most awesome force in the universe, Pop Murphy's Laws of Plumbing.

Pop Murphy's Laws of Plumbing

Most books describe pipes as if they were just inanimate objects. The truth is that pipes answer to a greater force, the one described on some heretofore lost tablets recently discovered buried beneath a mountain of plumbing parts at a local hardware store. The universal truth is that it's impossible to do any plumbing job without something going awry.

Pop Murphy's Laws of Plumbing are numbered to enable you to refer to each one easily throughout the book. The only antidote known for Pop Murphy's Laws of Plumbing is laughter—so apply this important remedy with carefree abandon. But don't forget, Pop Murphy, like rust, never sleeps!

The Official List of Pop Murphy's Laws of Plumbing (PMLP)

(These are *real*. I know; most have happened to me!)

Numero Uno — No matter how small the job, you will be missing one part—but only when the hardware store is closed.

#2 - The thought, "Just one more twist," instantly produces a cracked pipe or leaky thread.

#3 - Whenever a job calls for two pipe wrenches, only one is to be found.

#4 - Leaks will always wait for the most inopportune times—like just after you've put on your tux.

#5 - When you drop a threaded metal pipe, the pipe always manages to turn around in midair and land directly on the threads—damaging them beyond use.

#6 - With any part that has an arrow clearly marking the direction of the flow of water, you will install it backwards at least once.

#7 - Even though you know PVC glue is toxic and should not touch skin, you will decide to "skip the gloves *just this once* and be extra careful." Before you finish the word "careful," a dribble of glue will already be drying on your ungloved hand.

#8 - Pipe dope is the only correctly named plumbing part.

#9 - When a plastic pipe breaks, it will always break where the threads meet an expensive fitting so that the fitting is rendered useless.

#10 - Pipes are just the *Titanic* and water turned inside out. (Think about it!)

#11 - Be forewarned: Pop Murphy is, at this very moment, perfecting a gopher capable of gnawing through plastic PVC pipe—metal-chewing gophers will take a bit longer.

#12 - Pop Murphy was temporarily diverted from plumbing devices with the start-up of the Star Wars defense program. But now that he has trained the Pentagon in all his best management techniques, he has returned to designing the all-digital, all-confusing, microchipped, maxi-priced irrigation-controlling device.

Reality-Based Plumbing

Having 12 years of grubby hands-on experience on which to draw, I have distilled into this book what I've learned from thousands of hours of self-inflicted mistakes. Instead of explaining a number of different ways to design a drip system for each portion of your garden, I will outline the systems I've come to rely upon after years of messing around with many brands and types of gizmos and widgets. If this book helps you skip even 40 percent of the learning curve that I went through, I will feel it's worthwhile. No book can solve every irrigation question because all gardens may not match the book's examples. My intent is to explain the basic principles of design, familiarize you with the names of the best parts to use, provide a simple set of design guidelines, explain some tricks of the trade, and provide a list of resources for parts.

But, ah, the rewards awaiting the persistent budding student of "dripology." Upon finishing this book, you'll have a working drip irrigation system and you can endow yourself with an honorary Degree in Dripology (See **Appendix**) from that prestigious school of higher education—the School of Hard Knocks.

A Grounded Beginning

How long, how much and how widely and deeply you plan to water greatly affects the nuts and bolts of a drip system. Before you can begin to design your own system, it is important to know something about the soil in which you'll be planting and to understand some important principles about irrigation. While the following little discourse on soils and irrigation may seem like a digression, please bear with me and read through it before getting into the nitty-gritty of parts and gizmos.

1 Why Drip Irrigation Works — Anywhere

To understand how drip irrigation works, you need to know how your garden's soil affects the flow of water from the emitters (the small device which controls the water to a small trickle or dribble), how roots absorb moisture and nutrients, and how plants respond to wet and dry cycles.

The emitters release water very slowly and form a wet spot, mostly beneath the soil's surface. The shape of the moist area is affected differently by each type of garden soil and ranges from long and carrot-like to squat and beet-shaped. Knowing how your garden's soil affects an emitter's flow is important to planning your drip irrigation system.

Get to Know Your Soil

To test your soil, buy a single emitter, use an emitter punch (which cuts a tiny circular hole) to pierce a hole near the bottom of an empty plastic 1-gallon milk jug, and insert the emitter's barb. **(See Figure 1.)** Fill the jug with water and place it on a dry spot in the garden. After the milk jug is empty (which may take 24 hours or more because it's not under pressure), dig a small trench next to the emitter to see the shape of the wet spot in your soil. This is the most graphic way to understand how the water from each emitter you install will move laterally and vertically in your soil. Put the jug in several places in the garden to see how different soils affect the shape of the moist spot. Naturally, the width and depth of the wet spot and the degree of moistness will also depend upon the flow rate of the emitter and the length of watering time. Emitters with a higher rating, such as 2 or 4 gallons per hour (gph) will cause the wet spot to be wider than will 1/2- or 1-gph emitters.

Notice how the surface soil around the emitter is moist but not puddled, as the soil would be with the use of an oscillating sprinkler for the same amount of time. Also note that the wet spot you've exposed in the trench's wall is also moist, but not soggy. It is this moist-but-not-too-wet spot which is the key to drip irrigation's superiority over all other forms of irrigation. With proper timing, a drip irrigation system provides moisture to the soil with little puddling and without

overly saturating the pore spaces in the soil. This labyrinth of minute pore spaces helps the soil breathe. When a fairly aerobic condition in the soil is maintained by drip irrigation, the roots don't drown, the soil's beneficial bacteria can continue to release valuable nutrients and harmful anaerobic fungi don't easily proliferate.

The dry surface between the emitters' wet spots discourages weeds because dormant seeds don't have enough moisture to germinate. However, although reduced weeding is often touted as an important benefit of drip irrigation, this pertains only to arid climates. Any periodic summer rain will negate the effect of dry spots between plants by providing enough moisture to sprout weed seeds. Mulch, which is routinely used to hide a drip irrigation system, can be used to suppress any weeds which might germinate.

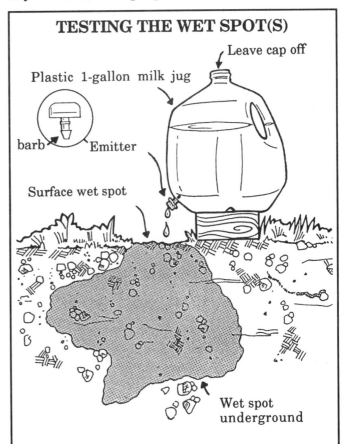

Figure 1 Use an emitter inserted into a plastic milk jug to learn about the flow of water in your soil. The wet spot will be larger when the emitter is pressurized by a complete drip irrigation system.

Where Roots Drink and Eat

It seems logical to many gardeners to water plants deeply, with infrequent but lengthy irrigations. Drip irrigation can be used for this kind of irrigation, but the roots of many plants, it turns out, don't really get most of their water and nutrients from the deep regions of the soil. In all the studies I've been able to find, the usual conclusion is that for the sake of quality growth, as opposed to sheer survival, the upper 1 or 2 feet of the soil accounts for over 50% of all the water a plant absorbs. **(See Figures 2 & 3.)** While many plants have roots deeper than 2 feet, these deeper roots exist mostly to stabilize the plant, absorb some micronutrients and help the plant survive droughts rather than to support an abundance of growth.

The biology behind this phenomenon is quite interesting. The upper layers of the soil are the most aerobic, with the highest population of air-loving bacteria and soil flora. As seen in **Figure 4**, the top 3 inches of the soil has nearly four and a half times more bacteria, almost eight and a half times more actinomycetes (tiny aerobic organisms which help decompose dead plant tissue), more than twice as many fungi and five times the algae of soil found 8 to 10 inches deep. These bacteria and flora are responsible for the decomposition of organic matter and the liberation of mineralized (unavailable) nutrients into a soluble form that the plant can absorb. They live near the soil's surface because they must have plenty of oxygen to fuel their activity.

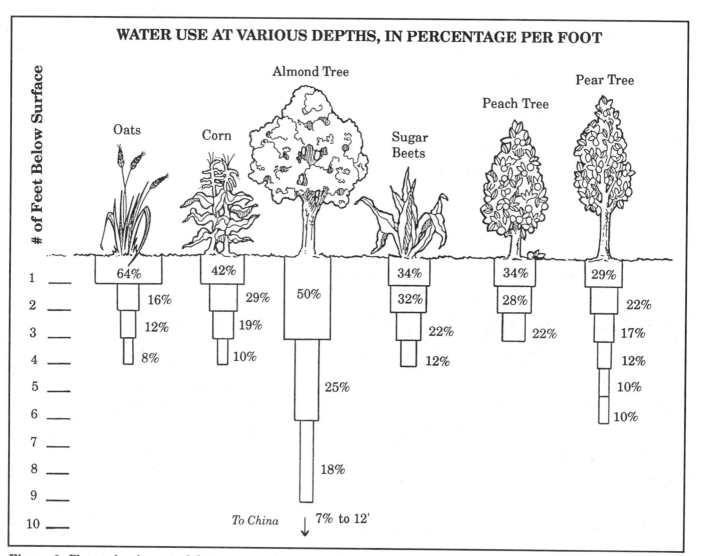

WATER USE AT VARIOUS DEPTHS, IN PERCENTAGE PER FOOT

Figure 2 Plants absorb most of their water and nutrients from the upper 1 to 2 feet of the soil. Each example is from a different independent study.

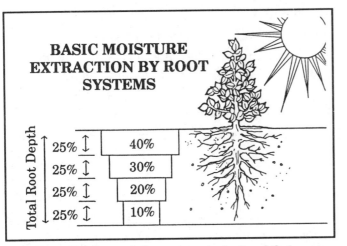

BASIC MOISTURE EXTRACTION BY ROOT SYSTEMS

Total Root Depth

25%	40%
25%	30%
25%	20%
25%	10%

Figure 3 The roots of plants get a majority of their water and nutrition for good growth and high yields from the upper 50% of their root zone. This illustration represents the average absorption by root depth.

The Wet and Dry Cycles of Plant Growth

A plant can absorb nutrients only in a water solution, and only when the soil is moist, not too wet or too dry. If the soil is too dry, nutrient uptake is inhibited because the soil life can't thrive and what little water is in the ground becomes tightly bound to the soil particles. Allowing the upper soil to dry out between infrequent irrigations means that nutrient uptake also "dries up." Then, when plenty of water is again supplied, the soil is saturated to the point that roots and air-loving soil life may be stressed or killed from *too much* water. Thus, nutrient uptake is inhibited by the lack of air. It also takes some time for the air-loving bacteria to re-populate a soil which is either too wet or too dry, so at either extreme there is a biological lag before the roots get their best meals. Infrequent, deep irrigations tend to produce two points in the watering cycle where the soil life is damaged enough to reduce or prevent growth—during the drying stage and when the soil is too wet.

Soils Often Aren't Very Deep

Another hidden assumption about deep watering concerns the depth of the fertile soil. There are places in the world where glacially deposited topsoil extends down for dozens of feet, but these are more the exception than the rule. If you have such a deep, loamy soil, then rejoice, but remember that the majority of

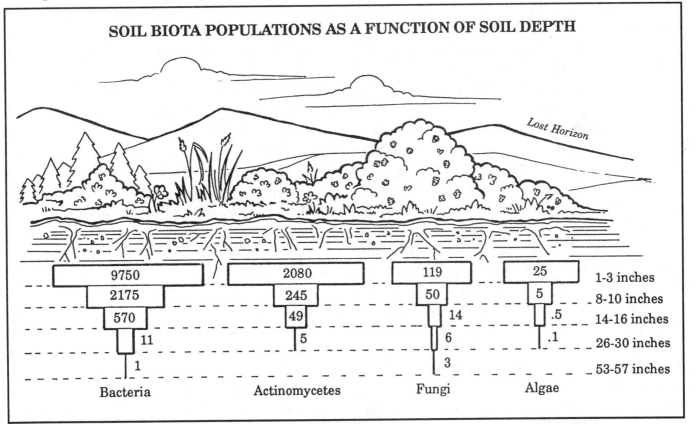

SOIL BIOTA POPULATIONS AS A FUNCTION OF SOIL DEPTH

Lost Horizon

Bacteria	Actinomycetes	Fungi	Algae	Depth
9750	2080	119	25	1-3 inches
2175	245	50	5	8-10 inches
570	49	14	.5	14-16 inches
11	5	6	.1	26-30 inches
1		3		53-57 inches

Figure 4 The upper levels of the soil are the most aerobic. The air-loving soil biota thrive near the soil's surface and are greatly responsible for the high availability of nutrients in the upper layers of the soil. (Numbers indicate individuals per zone.)

moisture and nutrient absorption still happens in the top 2 feet. Typically, most suburban yards have a very shallow layer of topsoil, if there's any left at all after construction. If, for example, there is a continuous layer of rich orange clay some 18 inches under the ground, then the 18-inch layer of topsoil is, for all practical purposes, the only place where the plant's roots will be feeding.

Most heavy clays, whatever the color—and dark blue and white-to pale-gray are the worst—are relatively worthless to feeding roots. While clayey soil may have plenty of nutrients, their availability is locked up in its tight, anaerobic structure and strong chemical bonds.

Frequent, Shallow Watering Is Best

For the sake of real growth and the absorption of moisture and nutrients, the loamy upper horizon is the important place to concentrate irrigation and fertilizers. Frequent waterings, if done in a way which avoids puddling or flooding of the soil, produce the best growth. In rainy areas, if watered just enough between periodic storms to maintain a moist-but-not-wet soil, your garden will produce higher vegetable yields or lush ornamental foliage. And the best tool for frequent, gentle watering is drip irrigation.

(This is not to suggest that drip irrigation can't be used for infrequent watering of plants hardened to a regimen of drought stress and minimal irrigation. Where water is very limited, drip irrigation is still the most efficient way to apply this precious resource. I live on a parcel of land where the normal summer drought has always been a problem. The fractured geology of the area makes it hard to find underground water, and most wells are poor producers. To conserve my water supply for essentials, I water portions of my landscape only three or four times during the entire six-month dry season. While this is far from ideal, my plants appear fairly healthy.)

Drip Irrigation Improves Yields and Growth—Everywhere

Research in many different climates and states invariably supports the benefits and cost-effectiveness of drip irrigation. Art Gaus, an extension horticulture specialist with the University of Missouri at Columbia, MO, has had a drip system in his personal garden for nine years. One summer, his bush watermelons with plastic mulch and a drip system produced 32 pounds in a 4- by 4-foot area, compared with 9-16 pounds in an area with conventional irrigation. He reckons a well-timed drip system "could mean a 100% increase in yields; during the droughts of 1980, '83, and '84 it meant the difference between having a crop or no crop at all." In a study of established pecan trees in Georgia, trees with drip irrigation had a 51% increase in yields. Michigan State University has documented a 30% yield increase in vegetable crops with drip irrigation, even in their humid, summer-rain climate.

I prefer frequent watering with small amounts of water, sort of like "topping off the tank." After the winter rains are over and the soil has reached an ideal moisture level, not too anaerobically wet and not too dry, the goal is to replace, as often as daily, exactly the amount of moisture lost due to evaporation from the soil and transpiration from the plant's leaves (called the evapotranspiration rate, or ET), plus an amount that represents enough extra water for gorgeous growth. *Arboriculture,* the preeminent text on growing and caring for trees, recommends frequent irrigation: "In contrast to other systems, drip irrigation must be frequent; waterings should occur daily or every two days during the main growing season . . the amount of water applied should equal water lost through evapotranspiration."

This is just an introductory look at the basics of soil and watering. Later, in the chapter titled When and How Long to Irrigate, I'll talk more about the specifics of timing. And in the chapter titled Drip Irrigation for Containers, more detailed guidelines for healthy watering will be discussed.

Now it's time to get into the hardware of drip irrigation—the gizmos, the widgets and gadgets. If drip irrigation seeems like a huge ugly bull, it's time, in the words of W. C. Fields, to "grab the bull by the tail and face the situation"! Most people's fears of drip irrigation gadgetry stem from a mild case of "gizmophobia." With a gentle introduction, some time to become familiar with the various parts and a little practice, most overcome their initial hesitations.

2 Basic Drip Irrigation Stuff

To paraphrase comedian George Carlin, "People love their 'stuff.' The more 'stuff' the better. But where are they gonna put all their 'stuff?' " Well, drip irrigation can add a lot more "stuff" to your yard. A poorly designed system can clutter your garden with cheap, breakable "stuff." A well-designed system keeps clutter to a minimum by maximizing the effectiveness of whatever bits and pieces you do wind up with and by hiding much of the system with an attractive mulch.

So, under the heading of absolutely necessary "stuff" for a good basic drip irrigation system are a *backflow preventer* to keep water in the hosing from siphoning back into your home's pipes; a fine-mesh *filter* so the small orifices in the emitters don't clog; a *pressure regulator* to keep the easily assembled fittings from blowing apart; and carefully placed lengths of small-diameter black hosing with emitters to control the flow of water. Important "stuff," but not too much, and all easily hidden or disguised.

The Main Assembly

At the beginning of any drip irrigation system, right where you attach it to your garden faucet, is the *main assembly* which consists of the backflow preventer, filter and pressure regulator. The main assembly services the drip irrigation hose and a number of branched lines of drip irrigation hose called *laterals*, or *sub-systems* (**See Figure 5.**)

The easiest way to begin installing a drip irrigation system is to use the existing garden faucet closest to your plantings. If you must add a faucet to the garden, be sure to use a solid brass one, not one of those shiny fake-chrome or plastic versions, which can so easily split.

All drip irrigation systems require the three above-mentioned parts of the main assembly in order to work effectively and safely. But many discounted drip irrigation kits, as featured at hardware and department stores, either don't include all these parts or offer only cheap, easily broken versions of one or more of the items. For good success and long-term performance, I recommend spending a bit more to get quality parts for a more reliable, longer-lasting system. In this way, you

may be able to get one up on Pop Murphy (see pages 4 and 5). I've broken and thrown away dozens of cheap parts over the years, so what follows is a list of my very specific recommendations based upon the successes and failures of years of installing drip irrigation systems for clients and in my own garden.

The Backflow Preventer

A backflow preventer (**See Figure 6**) is essential for the health of your family. Turning off any irrigation system can start a slight back-pressure suction, which can create a reverse siphon of water into the house's plumbing. This reverse siphon can suck dirt, mulch, bacteria, fungus, pesticides, herbicides, manure or fertilizer through the emitters, into the drip hose and back into your home's drinking water supply. It is important that a backflow preventer be installed at *every* faucet—whether a drip system is attached to that particular faucet or not.

There are two types of inexpensive backflow preventers—an atmospheric vacuum breaker and a check-valve. The atmospheric breaker creates an air gap to prevent the water in the garden hose or drip system from siphoning back into the drinking water. One drawback in the design of the atmospheric vacuum breaker is the amount of water wasted when water drains out of the drip hose and puddles up around the faucet each time the drip system is shut off. Because of the safety of the air gap, however, the atmospheric vacuum breaker is becoming the preferred code-required device for anti-siphon protection at faucets. The check-valve has a spring-loaded, gasket-rimmed plunger which quickly seals against a circular seat when the water flow reverses. A check-valve shuts off quickly and doesn't waste water, (compared with an atmospheric vacuum breaker), but pieces of organic matter, bits of dirt, or a buildup of algae slime, sand or tiny pebbles can keep the plunger from fully sealing. Thus, cautious enforcers of plumbing and building codes will not allow a check-valve alone as an anti-siphon device.

The best backflow prevention for a faucet assembly combines both an atmospheric vacuum

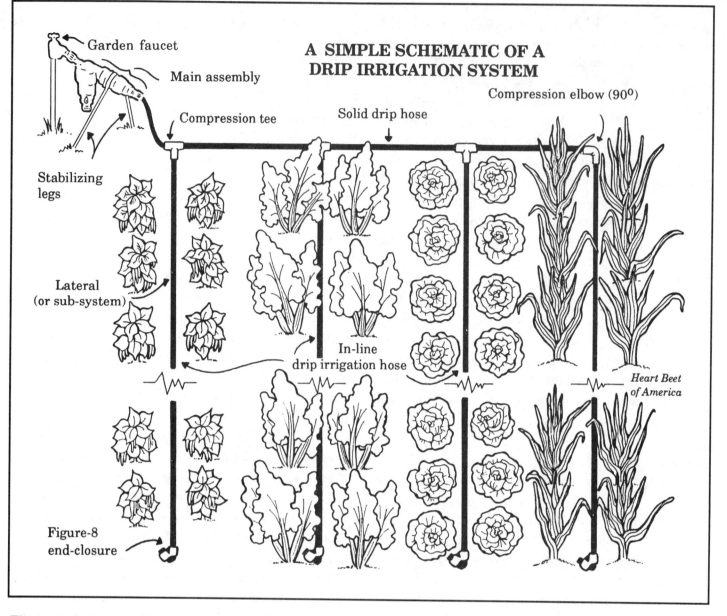

A SIMPLE SCHEMATIC OF A DRIP IRRIGATION SYSTEM

Garden faucet

Main assembly

Compression elbow (90°)

Compression tee

Solid drip hose

Stabilizing legs

Lateral (or sub-system)

In-line drip irrigation hose

Heart Beet of America

Figure-8 end-closure

Figure 5 A drip irrigation assembly attached to an existing faucet in the garden. Solid irrigation hose acts as the main supply line to the vegetable garden. Lines of in-line drip irrigation hose, with emitters built inside the hose, form the laterals along each row of vegetables.

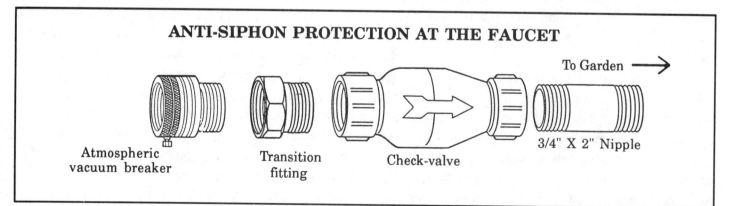

ANTI-SIPHON PROTECTION AT THE FAUCET

To Garden →

3/4" X 2" Nipple

Atmospheric vacuum breaker

Transition fitting

Check-valve

Figure 6 The combination of an atmospheric vacuum breaker and a check valve is the best way to protect your family from dirty water back-siphoning into the home's water supply. Be sure to point the arrow on the check-valve away from the faucet and toward your drip irrigation system. Make sure these two parts are at least 12 inches above the highest part of the drip system.

breaker and a check-valve to get the advantages of both (**See Figure 6**). A good-quality metal atmospheric vacuum breaker costs between $5 and $10. Don't go cheap and use a plastic one, as plastic can't take the stress of the rest of the main assembly hanging off the faucet. The atmospheric vacuum breaker comes with hose threads and is threaded directly onto the faucet before the all-metal hose Y-valve, which has two small ball-valves (**See Figure 23** in Chapter 3). The Y-valve allows both a drip irrigation main assembly and a spare hose for occasional watering tasks to be attached to the same faucet. The atmospheric vacuum breaker will also keep the garden hose from siphoning water back into the house. A check-valve costs $6 to $20, and it should be added in sequence after the atmospheric vacuum breaker. Use a check-valve that has a spring rated at 1/2 pound, so it will close quickly with any reverse flow. Be sure to note the arrow embossed on the check-valve's body or printed on the label. Don't forget Pop Murphy's Law of Plumbing (PMLP) #6 on page 4. The check-valve *must* be installed with the arrow pointing away from the faucet, toward the the drip irrigation hose. If you use a check-valve with a plastic body, don't tighten it too hard, as cracks may develop.

It is very important to make sure these two devices are *at least* 12 inches higher than any part of the drip irrigation system. If the supply tubing from the faucet travels uphill, the atmospheric vacuum breaker and check-valve must be installed 12 or more inches higher than the highest portion of the tubing.

The Filter

Without a good filter, the emitters can clog, rendering your investment of time and money useless. The filter can also be the Achilles heel of a drip irrigation system. Routine cleaning of the filter is essential to a properly functioning drip irrigation system. Filters which have to be taken apart to be cleaned are often ignored and forgotten. A quality filter which is easy to clean can guarantee easy, long-term maintenance. Therefore, for my money, the only filter worth getting is a Y-filter with a ball-valve at the end of the filter chamber for easy flushing (**See Figure 7.**) Priced from $12 to $20, a Y-filter is more expensive than an $8 tube-type filter but much easier to clean.

The diagram of a Y-filter in **Figure 8** shows

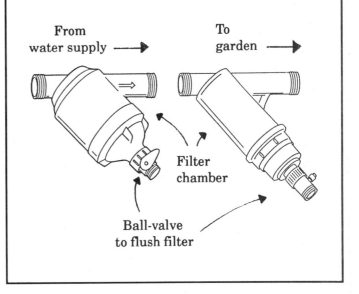

TWO TYPES OF Y-FILTERS

From water supply ➡

To garden ➡

Filter chamber

Ball-valve to flush filter

Figure 7 Two models of a Y-filter, the best filter to protect your emitters from clogging. The ball-valve allows for easy and rapid flushing of the filter without your having to take it apart.

how the water passes from the inside of the screen cylinder to the screen's exterior and on to the drip line. The ball-valve built into the bottom of the filter chamber allows you to flush out dirt by simply opening the ball-valve and letting the water run for a few minutes. The water passes through the interior of the screen at great velocity and flushes out the trapped silt and sand. Because a strong blast of water is best for cleaning the filter, the filter *must* be plumbed into the main assembly *before* the pressure regulator so it can get full, unrestricted water-pressure. The second advantage of the Y-filter is its capacity to pass water. Y-filters have a much larger internal filter surface than tube-type filters, allowing for more water to pass per minute to supply a greater length of drip hose. The typical 3/4-inch Y-filter with a 150-mesh screen (able to filter out particles down to 106 microns—the larger the mesh number on a screen, the smaller the particle which is trapped) passes up to 660 gph, or over four times as much water as a tube-type filter. Thus, while Y-filters appear to be more expensive, they are far more efficient and cost-effective; a single $15 Y-filter does the cleansing work of $35 worth of the tube types.

Be sure to purchase a Y-filter with a metal

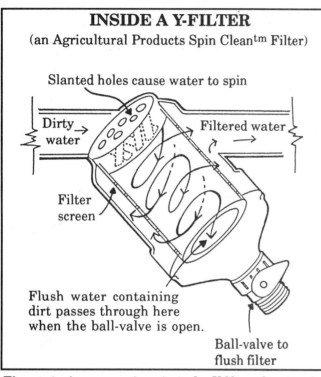

INSIDE A Y-FILTER
(an Agricultural Products Spin Clean™ Filter)

Slanted holes cause water to spin

Dirty water

Filtered water

Filter screen

Flush water containing dirt passes through here when the ball-valve is open.

Ball-valve to flush filter

Figure 8 A cross-section view of a Y-filter, showing the swirling vortex which carries dirt or sediment to the bottom of the filter's metal screen cylinder for easy purging. Opening the ball-valve allows the water pressure to flush the accumulated dirt out of the filter. The slanted holes leading into the filter, which cause the vortex, are a patented design of Agricultural Products.

screen filter cylinder, not one made of plastic mesh. The metal screen can be scrubbed periodically, usually once a season, with a toothbrush and a solution of 10% bleach to remove any algae or scum buildup. Plastic mesh screens won't stand up to such scrubbings.

The Pressure Regulator

Every drip irrigation system also needs a pressure regulator to prevent the simple nonthreaded drip irrigation parts from blowing apart. A regulator which keeps the pressure below 25 pounds per square inch (psi) is often sufficient, but with some drip system parts, an even lower psi is required. Regulators are available with 10-, 15-, 20-, and 25-psi ratings. Porous hose (see below) must be kept below 10 psi so the fittings don't leak. Laser™ drip tubing (1/4- or 3/8-inch tubing with small slits cut at regular 6- or 12-inch intervals by a laser) is especially useful in large pots or in small vegetable and flower beds, but it can work correctly only at 10 to 15 psi.

Good-quality pressure regulators, the ones I use, are made by Senninger™ **(See Figure 9)** are sold for a specific range of minimum and maximum flow. The regulator I use most often is the Senninger 25-psi low-flow model markied .1-8 gallons per minute (gpm). This means that the regulator only functions correctly when the flow rate is greater than six gallons per hour (.1 gpm times 60 minutes per hour) and less than 480 gallons per hour (8 gpm times 60 minutes). At higher or lower flow rates, the regulator will not control the pressure effectively.

Here's how to find the flow rate for each of your drip lines in order to purchase the correct regulator: Determine the flow of your system. Count the number of emitters you have, or plan to have, on each line. Multiply by the flow rate of the emitters to get the total flow rate. (If you have emitters with various flow rates, total the flow rates for all the emitters.) Make sure this total falls within the range of the pressure regulator you've chosen. For example, if you have a total of 37 emitters along a length of drip hose and each emitter passes 4 gph, the total flow is 148 gph or 2.5 gpm. This is well within the flow rate of the Senninger low-flow model: 6 to 480 gph, or .1 to 8 gpm. As another example, using in-line emitter tubing with 1/2-gph emitters every 12 inches, the Senninger low-flow pressure regulator passes enough water to use 12 to 960 feet of tubing. Or, the same regulator could supply between 3 and 240 emitters rated at 2 gph.

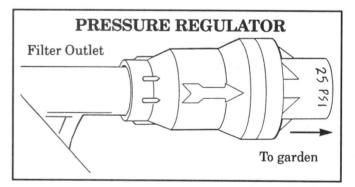

PRESSURE REGULATOR

Filter Outlet

25 PSI

To garden

Figure 9 A pressure regulator is required to protect the easy-to-use drip irrigation fittings from blowing apart. Be sure the arrow points to the drip system. The pressure should be between 10 and 25 pounds per square inch (psi), depending upon the type of system. Regular house water pressure ranges between 40 and 60 psi. The Senninger™ pressure regulator comes with either hose or iron pipe threads and with female or male threads. This one has female iron pipe threads (fipt) to make it easy to assemble with the filter's male iron pipe threads (mipt).

Different Ways to Drip Your Landscape

There are three types of hardware, or systems, which can be used in the garden or landscape—*porous hose*, drip irrigation hose with *punched-in emitters,* and *in-line emitter tubing*. Each has various benefits and limitations.

Note: You may have heard of "spaghetti" tubing, thin 1/4- or 1/8-inch tubing which is used to run from a 1/2-inch solid drip hose to each plant. I think "spaghetti" is a good description—the stuff really gets tangled. Also, it is practically impossible to weed or mulch around and can easily snap off of the 1/2-inch drip hose. Personally, I never use "spaghetti" in the landscape, reserving it for limited use with container plants only (to be discussed later).

Porous Hose

Porous hose, also called soaker hose, oozes water through a labyrinth of minute channels in the hose and out of its entire surface. The soil is watered along the entire length of the hose and out from it to a width which depends upon both the type of soil being irrigated and how long the water is left on. Porous hose is made of either used tires or new tire trimmings combined with low-density polyethylene. The fittings sold to install porous hose are inserted inside the hose and will either leak or come apart if the pressure is too high, so be sure to use a pressure regulator that will keep the pressure below 10 pounds-per-square-inch (psi). Porous hose is often sold without a filter, but you should install one to prevent possible clogging. Even with city water, a 200-mesh filter (which traps all particles 74 microns or larger) will help protect your investment.

The benefits of porous hose include: it is sturdy, is readily available through mail order companies, is very easy to hook up to your existing faucet, is simple to snake around your existing plantings, can be buried in the soil, drains itself for freeze protection, and works with pressures as low as 10 psi.

Drawbacks include: the 5/8-inch version is rather bulky and ugly if left uncovered; the metal hose clamps (often sold to guarantee that the insert fittings don't leak) are expensive, and some eventually rust; and unchlorinated well water can produce an algae slime which eventually seals the pores on the interior of the tubing. While resistant to rodents, it's not gopher-proof; flow rates can vary considerably from the beginning to the end of a line; and, to avoid differences in flow, it can be used only on flat ground.

Be wary, by the way, of manufacturers' guidelines on how far apart to space each length of porous hose. I know an employee of a mail-order company which carries porous hose who followed the company's own guidelines for spacing the hose below the ground for a lawn. His lawn ended up striped a charming green-brown-green-brown. Before you bury any porous hose, do a test run to see how wide the wet spot will be for a given length of irrigation.

Emitters

Emitters come in dozens of configurations (See Figure 10), usually in flow rates of 1/2, 1, 2 and 4 gph, and are inserted into 1/2-inch (18mm) polyethylene drip irrigation hose. Emitters are commonly used for what I call point-source drip irrigation, in which only one or two emitters are placed near the stem of each plant. There are both regular and pressure-compensating emitters. With regular emitters, the flow rate from one end to another (on lines totaling more than 100 feet or from one elevation to another) can vary considerably. If the total up-and-down change in your garden's topography is more than 20 feet, you must use pressure-compensating emitters, which are designed to put out almost the same amount of water regardless of topography and length. Even though pressure-compensating emitters can cost 30% to 100% more, when I choose punched-in emitters, I always use the pressure compensating type to make sure the flow is consistent throughout the system.

Limitations with the use of pressure-compensating emitters are that the parts are not as widely available by mail order as porous hose, the stem of the emitter will get brittle with age and can easily snap off during weeding, hoeing or mulching and, in a large garden, installing hundreds of emitters has been known to injure a wrist or shoulder.

In-Line Emitters

In-line emitters are the least well known drip

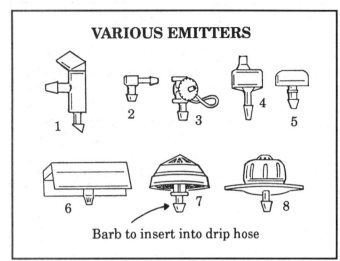

VARIOUS EMITTERS

Barb to insert into drip hose

Figure 10 Emitters control the flow of water from a drip irrigation hose and come in many sizes, shapes and flow rates. Each emitter has a barbed end so it can be inserted in a hole punched into the solid drip irrigation hose. All of these emitters are noncompensating, unless otherwise noted. **1**—Hardie DBK™, **2**—Microflapper™ (pressure compensating), **3**—Spot Vortex™, **4**—Rainbird Rainbug™ (pressure compensating), **5**—Netafim Button™, **6**—Olsen Turbo Flow™, **7**—Netafim RAM™ (pressure compensating) and **8**—Spot Scrubbler™ Emitter—adjusts from 0 to 10 gph.

irrigation technology, but offer, in my experience, the best mix of efficiency, ease of installation and resistance to clogging and leaking. The tubing is nearly 1/2 inch in diameter (16mm, as opposed to 18mm for "regular" 1/2-inch drip hose) and comes with an emitter preinstalled inside the tubing at regular intervals. These internal emitters utilize what is known as a "tortuous path" **(See Figure 11)**; the water must pass through a labyrinth of right-angled channels inside the emitter before exiting via a hole much larger than that of a typical punched-in emitter. The tortuous path causes the water to form a continous vortex, a kind of horizontal tornado which keeps any sediment, sand or silt in suspension so it won't settle out and clog the emitter. I've used in-line emitters for 10 years, and even with well water full of soluble iron oxide, (notorious for clogging regular punched-in emitters) I've found only a few clogged emitters in several thousand feet of tubing. In-line emitter tubing moistens the soil the entire length of the line but slightly below the surface, where the bulbous-shaped wet spots come together to form one nearly continuous moist zone. **(See Figure 12.)** The emitters come pre-installed in tubing with 12-, 18-, 24-, and 36-inch spacings, but this type is most commonly

sold to gardeners in the 12-inch and 18-inch intervals. The emitters inside the hose are rated to dispense either 1/2 or 1- gph (actually .6 and .92 gph). Newer versions include tubing with pressure-compensating emitters at the same intervals as in the noncompensating in-line tubing and with 1/2- or 1-gph flow rates. **(See Figure 13.)** I always use the kind with 1/2 gph emitters on 12-inch centers because they will irrigate, depending upon how long the system is left on, both sandy and clayey soils.

The benefits of in-line pressure-compensating emitters are that they are easy to install, simple to snake around your existing plantings, suffer less clogging than porous tubing and some punched-in emitters, work at the greatest range of pressures (7-25 psi), provide consistent rates of irrigation without regard to slope or length and have no external parts to snap off, their connectors or fittings don't leak, and the connectors, either compression fittings or Spin Loc™ fittings, seal better than metal hose clamps with porous hose.

The drawbacks are few: they are not recommended for plants placed far apart and at odd intervals; they can't turn in as sharp a radius as porous tubing and

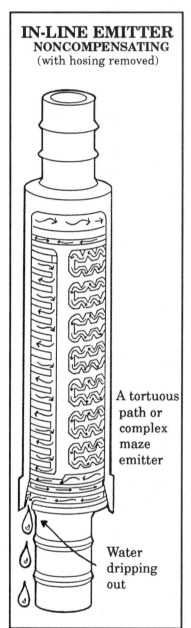

IN-LINE EMITTER
NONCOMPENSATING
(with hosing removed)

A tortuous path or complex maze emitter

Water dripping out

Figure 11 The in-line emitter is built inside the drip irrigation hose. This type has a complex path for the water to follow, known as a "tortuous path," which regulates the flow and helps keep the emitter unclogged. This is *not* pressure compensating.

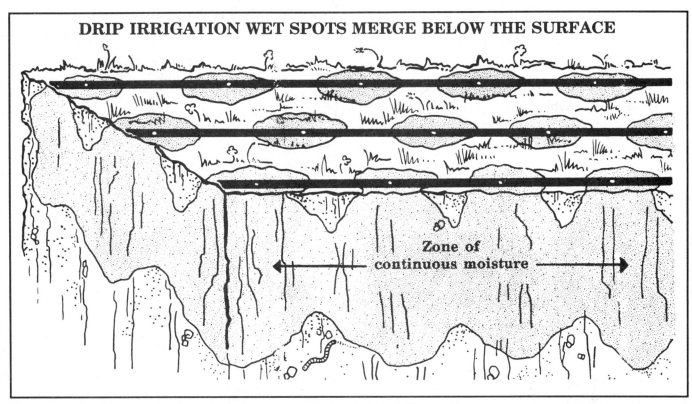

DRIP IRRIGATION WET SPOTS MERGE BELOW THE SURFACE

Zone of continuous moisture

Figure 12 This illustration shows how the wet spots beneath each in-line emitter merge to form one continuous zone of moisture. The soil for the entire length of the in-line tubing is moist some 4 to 6 inches beneath the surface, depending upon the soil type. Sandier soils require tubing with the emitters pre-installed 12 inches apart, while heavier, clayey soils need emitters only every 18 or 24 inches.

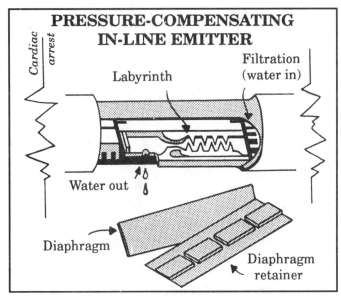

PRESSURE-COMPENSATING IN-LINE EMITTER

Cardiac arrest

Labyrinth

Filtration (water in)

Water out

Diaphragm

Diaphragm retainer

Figure 13 This is a cross-section drawing of a single emitter inside pressure-compensating in-line emitter tubing. The tortuous path is much shorter than in the noncompensating in-line tubing. The flexible diaphragm handles both pressure regulation and the flushing of any particles which might clog the emitter.

they aren't carried by very many mail-order or retail outlets. (See **Suppliers and Resources** near the end of the book.)

Flow Rates Compared

The main assembly represents a major cost in any drip irrigation system. The total flow rate of each type of tubing or hose has a direct influence on the cost of main assemblies. The lower the flow rate of your water system, the more main assemblies you'll need. With porous hose, each valve can reliably serve up to a total of 200 feet of tubing without a substantial reduction in water flow (with a flow rate of 160 gph at 10 psi). A main assembly or valve, servicing drip hose with punched-in emitters, can supply at least 200 feet of drip hose, for a maximum flow of 240 gph. With in-line pressure-compensating tubing, a single main assembly or valve can support up to 326 linear feet with 12-inch spacing and pressure-compensating emitters rated at 1/2 gph (for a total flow rate of 163 gph at 25 psi) or 503 feet with 18-inch spacings and 584 feet with

24-inch spacing. One-gph in-line pressure-compensating emitter tubing with the same 25 psi specifications can run 248 feet with 12-inch spacings, or 382 feet on 20-inch centers. (See the Appendix for the flow rates of various in-line emitter tubings.) This means that in-line tubing, at the lowest flow rates of 1/2-gph emitters on 12-inch centers, allows you to install at least one-third more length of hose per main assembly than either drip hose with punched-in emitters or porous hose. Therefore, the use of in-line tubing means you spend one-third less on main assemblies than you would for drip hose with punched-in emitters and porous hose. With more distant spacings between emitters, the savings are even greater.

Punched-in emitters do have the advantage of being able to allow more than 200 linear feet of total drip hose per valve, up to 1000 feet, providing the total amount of water emitted is less than 110 gph—such as 220 emitters rated at 1/2-gph each or 55 2-gph emitters. The limit of a total flow of no more than 110 gph is to compensate for the reduction in flow caused by the friction of water moving through the extra length of hose. In this situation, with more than 200 linear feet of tubing, you must use pressure-compensating emitters so that all emitters pass the same amount of water, regardless of the length of the hose.

Miscellaneous Stuff

Finally, there's a wide range of miscellaneous gizmos and widgets to personalize your system and make it more fun to work with. **Figures 14, 15, 16 &17** show various stakes, spaghetti-holders for container plantings, misters, transfer barbs, "goof plugs," spaghetti right-angle barbs and emitters. I'll discuss each of these gizmos later in the sections that apply to their use. Tubing mistakes can easily be corrected with "goof plugs," (small solid barbed plugs used to patch holes in drip hose) or by splicing in new tubing.

All drip irrigation parts are either hand-threaded and hand-tightened or easily inserted into each other with compression fittings, Spin Loc™ fittings or insert fittings. **(See Figure 18.)** One advantage to all drip fittings is that no messy, smelly, toxic glues are required.

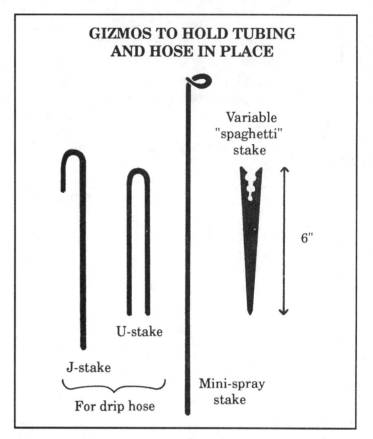

Figure 14 Various stakes and holders to keep drip irrigation hose and "spaghetti" tubing in place.

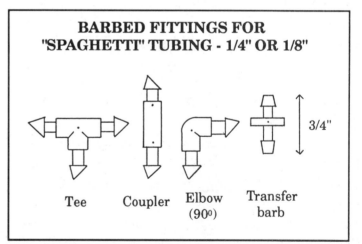

Figure 15 "Spaghetti" tubing is joined together with these barbed fittings. No messy glue is required. The transfer barb is used to make the transition fron 1/2-inch drip hose to "spaghetti" tubing.

With insert fittings, the drip hose fits *over* the barbed male portion of the fitting. And, depending upon the pressure rating on the regulator, hose clamps go around the outside of the drip hose to prevent leaks. Some people say the insert fittings don't need a hose clamp if the pressure is kept below 10 psi. I would never trust an insert fitting without hose clamps. The greater the water pressure, the weaker an insert fitting becomes. Should an insert fitting fail, the water will flow out at the rate of three to five gallons-per-*minute* instead of 1/2 to 2 gallons-per-*hour*.

With compression fittings, the drip hose fits snugly *inside* the female orifice of the fitting. **(See Figure 19.)** Unlike insert fittings, compression fittings actually seal better as the pressure increases, up to a point. At pressures above 25 psi, most likely at or above 50 or 60 psi, the fittings may split apart. But no drip irrigation system should be operated above 25 psi. I've always preferred compression fittings because they are much less expensive than insert fittings with the appropriate hose clamps and they are quick and easy to install—for me. Other people I've worked with have a difficult time inserting the drip irrigation hose into compression fittings. This is especially true with the pressure compensating in-line emitter tubing which has an outside diameter (O.D. .67 inch) slightly larger than the inside diameter of the green-ringed compression fittings (I.D. .63 inch). I've always been able to grab the drip hose close to its end and wiggle it into the female barbed opening. Sometimes, when other people are having problems putting the two together, I have them bring a thermos of hot water into the garden. If an inch or two of the end of the drip hose is stuck into the hot water, it softens up enough to slide into the fitting. Some folks, even with wiggling, hot water, well-placed curses and repeated attempts, just can't get the hang of joining the two parts together.

The Spin Loc fittings are for those who haven't been able to master the use of compression fittings with in-line emitter tubing. These fittings have a smooth nipple which has a blue O-ring gasket around the outside *and* a female-threaded tightening ring which threads over the drip hose. You simply pull back the tightening ring, insert the nipple into the hose, pull the ring down over the end of the drip hose and thread the ring onto the drip hose. The threads of the tightening ring cut a shallow set of grooves into the soft hose and secure the inside of the drip hose against the blue O-ring. Unlike compression fittings, these fittings are easily reused. Spin Loc fittings cost nearly twice as much as the equivalent compression fitting. Spin Loc fittings offer one very nice advantage over compression

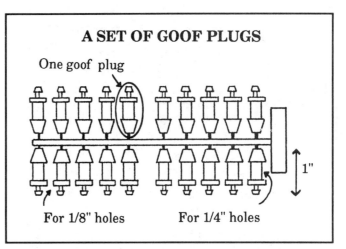

Figure 16 Most people make a few mistakes. These are goof plugs, for repairing mistakes when using "spaghetti" tubing and transfer barbs. One end of each of these goof plugs fits the holes of either 1/4- or 1/8- inch "spaghetti" fittings and emitter barbs.

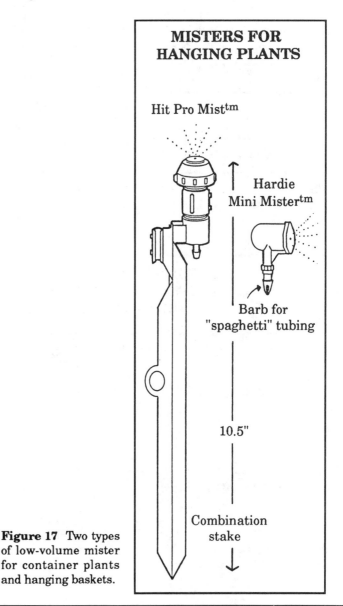

Figure 17 Two types of low-volume mister for container plants and hanging baskets.

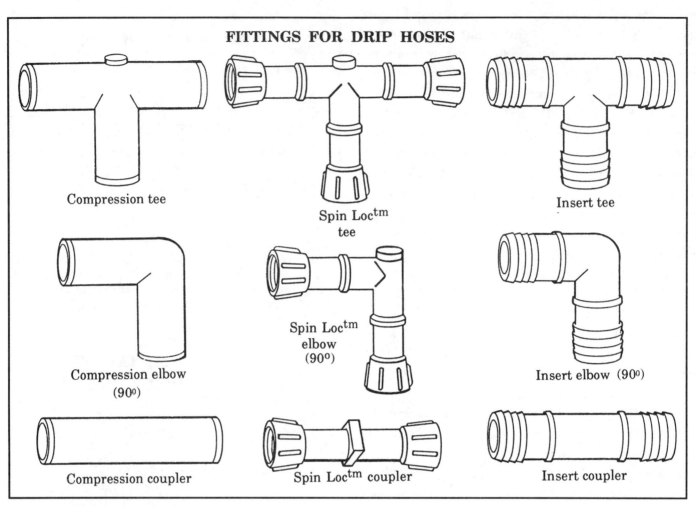

FITTINGS FOR DRIP HOSES

Compression tee

Spin Loc^{tm} tee

Insert tee

Compression elbow (90º)

Spin Loc^{tm} elbow (90º)

Insert elbow (90º)

Compression coupler

Spin Loc^{tm} coupler

Insert coupler

Figure 18 Various compression and insert (barbed) fittings for 1/2-inch drip irrigation hose. Each is easy to install and requires no messy glues, tools or threading.

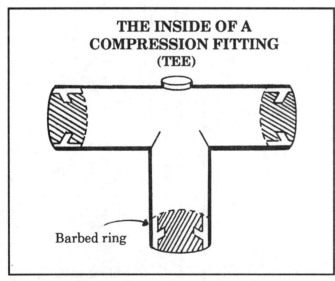

THE INSIDE OF A COMPRESSION FITTING
(TEE)

Barbed ring

Figure 19 The inside of this compression tee shows how the drip irrigation hose can be inserted but the barbed ring keeps it from being pulled or pushed back out.

fittings: the same Spin Loc fitting works with standard 1/2-inch drip hose (O.D. .708 inch, blue ring) *and* with in-line emitter hose (O.D. .67 inch, green ring). This means you won't have to stock two types of compression fitting. As this book is being written, I've just begun to test Spin Loc fittings in my garden—so far, they're performing fine. What I'm curious about is their ability to withstand long-term abuse and neglect, long freezing spells, the heat of summer and the changes of the seasons.

3 Your First Drip Irrigation Project

If you live where there is a municipal water system, you probably won't have a supply problem, but you should know the water pressure at each existing garden faucet. **(See Figure 20.)** You can buy a simple pressure gauge to hook up to the faucets and test the existing pressure in pounds per square inch (psi). **(See Figure 21** for an example of a pressure tester). Keep a written record of the psi for each faucet.

Checking Your Water Supply

People with wells should check their records for the flow-rate, or pumping test, that is usually made after installation. The important figure for drip irrigation purposes is the sustained pumping rate, in gph or gpm. This is vital to know in order to ascertain the maximum flow rate for each drip irrigation system. If you have a storage tank, this information isn't very important, but knowing your well's performance record will help you maintain a conscientious program of water management. Also check your records for any indication of dissolved iron and calcium; these minerals can clog certain emitters. If, for example, your well produces

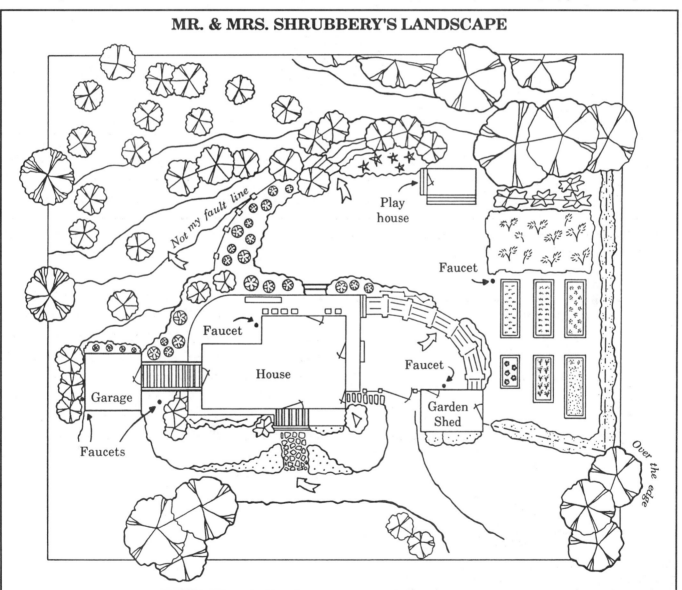

MR. & MRS. SHRUBBERY'S LANDSCAPE

Figure 20 Mr. and Mrs. Shrubbery live in a nice house with extensive landscaping. As luck would have it, they have installed every type of drip irrigation system mentioned in this book. They've graciously allowed me to copy the landscape plans for their yard. We'll look at various parts of their yard as examples of good drip irrigation design.

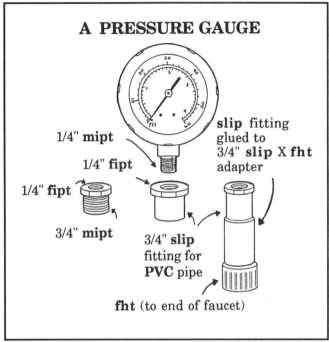

A PRESSURE GAUGE

1/4" **mipt**

1/4" **fipt**

1/4" **fipt**

3/4" **mipt**

3/4" **slip** fitting for **PVC** pipe

slip fitting glued to 3/4" **slip** X **fht** adapter

fht (to end of faucet)

Figure 21 A pressure gauge will tell you the pressure in pounds per square inch at each existing faucet. Some of these adapters may be needed to attach the gauge to your fixture.

iron-rich water, that fact greatly affects your choice of emitters. Those who use water from wells should also test and make a written record of the psi for each existing faucet.

We find the main assembly in the Shrubbery's yard at an existing faucet near the perennial flower bed **(See Figure 22)**.

Controlling Your Drip System

The most reliable way to control a drip system is to turn the faucet on and off manually. With a twist of the wrist, there's very little to go wrong. For your first drip irrigation system, stick with the existing faucet as the "controller."

The Sexual Anatomy of Plumbing Parts (Rated NC-17)

At this point, it's time for a polite reminder that plumbing parts are much like human parts— the protruding round parts stick into the hollow, tube-like parts. Therefore, the sticking-out parts with pipe-

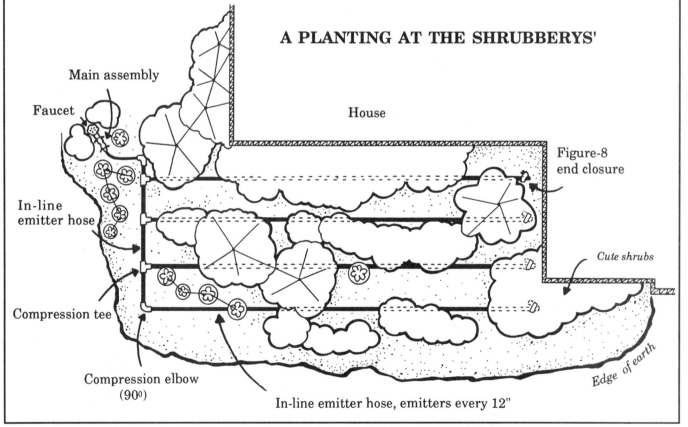

A PLANTING AT THE SHRUBBERYS'

Main assembly

Faucet

House

Figure-8 end closure

In-line emitter hose

Cute shrubs

Compression tee

Edge of earth

Compression elbow (90⁰)

In-line emitter hose, emitters every 12"

Figure 22 This view of the ornamental planting by the Shrubberys' house shows where the main assembly is located and the parallel lines of in-line drip irrigation hose. The mulch was removed to show the hosing. Even though the foliage doesn't cover all of the ground, the roots permeate the entire area and absorb moisture from all the emitters in the in-line hosing.

threads on the outside are the male pipe parts, and the hollow tubes with threads on the inside are the female parts. The male parts are labeled "mipt," which stands for male iron pipe-thread. The female ones are labeled "fipt"--female iron pipe-threads. Furthermore, some of these parts are made to be attached to a faucet and have unique threads called "hose threads." Thus, the end of a hose has a male hose thread (mht) fitting. The end that attaches to the faucet has a female hose thread (fht) fitting. A length of pipe with mipt threads on both ends is called a nipple. Pipe nipples come in many lengths, from ones only 3/4 inches long (called close

nipples) to 12 inches, 36 inches and even 48 inches, and many sizes in between.

The collection of parts for a main assembly shown in **Figures 23 & 24** is based upon years of trying many different manufacturers' products; these particular parts are the best I've worked with. This is a good setup for the drip irrigation novice because it's so easy to install. Another advantage is the high visibility of the assembly--it's easy to monitor, check for leaks, and disassemble for winter storage.

In the illustration, each fixture is referred to by its generic name, such as "check-valve," and by its

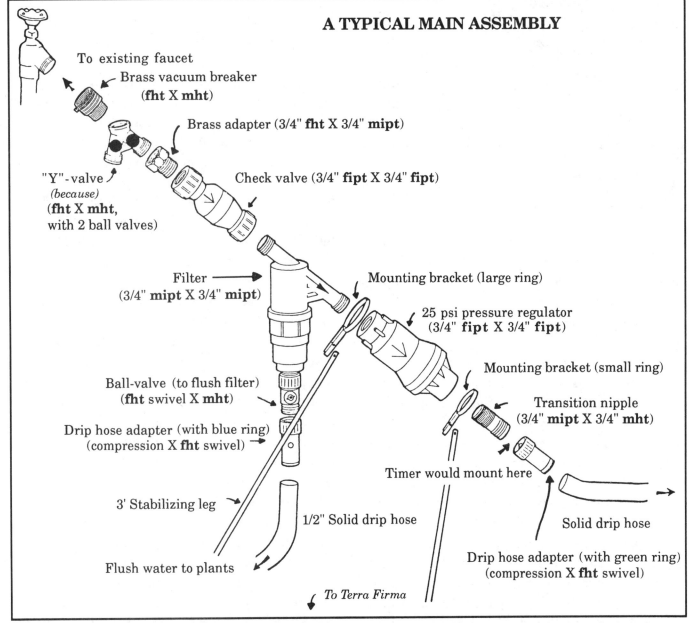

A TYPICAL MAIN ASSEMBLY

To existing faucet
Brass vacuum breaker
(**fht** X **mht**)

Brass adapter (3/4" **fht** X 3/4" **mipt**)

Check valve (3/4" **fipt** X 3/4" **fipt**)

"Y"-valve
(because)
(**fht** X **mht**,
with 2 ball valves)

Filter
(3/4" **mipt** X 3/4" **mipt**)

Mounting bracket (large ring)

25 psi pressure regulator
(3/4" **fipt** X 3/4" **fipt**)

Mounting bracket (small ring)

Ball-valve (to flush filter)
(**fht** swivel X **mht**)

Transition nipple
(3/4" **mipt** X 3/4" **mht**)

Drip hose adapter (with blue ring)
(compression X **fht** swivel)

3' Stabilizing leg

Timer would mount here

Solid drip hose

1/2" Solid drip hose

Flush water to plants

Drip hose adapter (with green ring)
(compression X **fht** swivel)

To Terra Firma

Figure 23 All the parts of a typical main assembly. The atmospheric vacuum breaker and check-valve keep the home's water pure, the filter protects the emitters and the pressure regulator protects the drip irrigation fittings from stress and breakage.

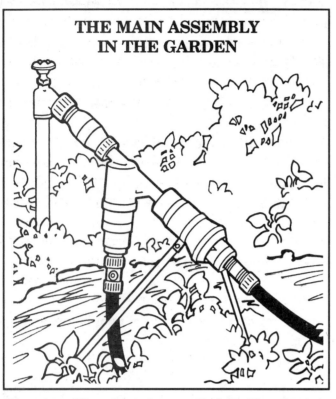

THE MAIN ASSEMBLY IN THE GARDEN

Figure 24 What the main assembly looks like after it is put together, attached to a garden faucet and mounted on protective legs. The tripod legs are essential to shield the main assembly from stress and breakage.

accurate plumbing moniker, such as "3/4-inch check-valve, fipt X fipt." The "fipt X fipt" code isn't some kind of esoteric plumbing math; it means "female iron pipe-thread *by* female iron pipe-thread," or, more simply, female iron pipe-threads on both ends of the check-valve.

Since some of the parts in the main assembly have hose-threads and some have pipe-threads, you'll need some adapters, or transition fittings, to join everything together. For example, in **Figure 23**, the transition nipple converts the female pipe-threads of the pressure regulator into male *hose*-threads to fit into the female hose-threads on the swivel of the drip-hose adapter. Be sure to keep track of which parts have hose-versus pipe- threads. It's hard to spot the difference, but hose-threads are more coarse and there are fewer per inch. You might want to mark the hose-threaded parts with a touch of paint or marker pen when you buy them. To try and join iron pipe-threads with hose-threads will ruin the threads of at least one part and produce a leaky union.

Putting the Assembly Together

Now, to assemble the assembly. You may want to loosely assemble all the parts first to become familiar with the fittings and their sequence, then take apart the assembly and lay out the parts in the same sequence. You may also want to have someone read the following instructions out loud as you put everything together. These instructions assume you already have an atmospheric vacuum breaker and a Y-valve attached to the garden faucet. The main assembly will attach to the faucet (mht) directly, but it's not recommended that you skip the atmospheric vacuum breaker.

1—To the mipt end of the brass adapter, add some Teflon™ pipe dope to make it easier to thread the parts together and to seal the threads against tiny leaks.

(Now, keeping in mind PMLP #8, a little digression about covering iron pipe-threads. Old-fashioned pipe dope is sticky gray stuff which tends to get all over your hands and sometimes dries out after it's applied—resulting in the inside-out *Titanic* effect. Many plumbers claim the solvents in traditional pipe dope are corrosive to modern plastic fittings. Teflon pipe dope, on the other hand, duplicates the slickness of a Teflon frying pan to make it easy to thread two parts together, lubricates the parts for an easy fit, appears to stay somewhat flexible for a permanent seal and is much easier to work with than the Teflon pipe tape discussed below. But Teflon pipe dope still manages to adhere to hands, pants, shirts and tools like balloons to a wool sweater. There is still much debate among plumbers as to the real or imagined damage from either pipe dope on plastic fittings. Teflon pipe *tape* is a thin white material which is wrapped around male threads and doesn't have any corrosive compounds. Pipe-threads seal up quite well with Teflon tape, in spite of the ease with which the parts go together. But it is a real hassle to work with out in the landscape. The gossamer-like tape blows around, knots up into useless balls, twists together just as you're about to apply it to the threads and picks up gritty particles like a magnet—making it much harder to get a good seal. If I'm prefabricating a main assembly indoors, at the workbench, I reach for the Teflon tape. Outdoors, I use the Teflon pipe dope instead of the bothersome tape. In the end, the choice is yours.)

2—Next, apply the check-valve until it is finger-tight. (In order to circumvent PMLP #6, be sure to note which direction the embossed arrow on the check-valve is facing. The arrow should face *away* from the faucet, and toward the drip hose.)

3—Use a pipe wrench to secure the brass adapter and another pipe wrench to twist the check-valve a one-quarter turn—don't tighten any further as you may manifest PMLP # 2 and crack the check-valve.

4—Now apply Telfon pipe dope to the mipt inlet part of the Y-filter.

5—After finger-tightening the filter, hold the check-valve securely with a pipe wrench and rotate the filter by hand *only* one-quarter of a revolution (to avoid cracking the filter's housing).

6—Next, slip the larger of the two mounting brackets onto the fatter end (the intake) of the pressure regulator. (The two mounting brackets can be purchased at a hardware store; they're the same as the galvanized fittings used to attach chain-link fencing to its metal posts.)

7—Now more pipe dope goes onto the mipt outlet of the filter.

8—Add the pressure regulator and finger-tighten it, then rotate the pressure regulator another one-quarter revolution. Be sure to use the model of pressure regulator which is designed to handle the volume of water the system will pass. In most cases, I use a Senninger 25-psi low-flow regulator, rated at 1/10 to 8 gpm (or 6 to 480 gph.)

9—Now you can add the transition nipple. With this nipple, you need only apply pipe dope to the mipt side and hand-tighten it into the female end of the pressure regulator. The mht (male hose-thread) end of the transition nipple allows you to add a swivel X compression fitting, also called a drip-hose adapter, to the end of the main assembly. One end of the drip-hose adapter has female hose-threads to attach to the end of the main assembly; the other has a special quick-couple orifice called a compression fitting. Compression fittings **(See Figure 19)** work by means of an internal ringed barb; the hose can be easily inserted into the fitting's hole, but the barb prevents its removal. To insert the drip hose into the compression fitting, moisten the end

of the cut hose and, holding the hose within 2 inches of its end, wiggle it into the hole until the barb pinches around the tubing.

10—Because the swivel end of the drip-hose adapter has a hose gasket, you can skip using pipe dope on the transition nipple and just hand-twist the swivel onto the mht end.

11- Now attach the brass adapter, at the beginning of the main assembly, to the hose-threaded Y-valve that's already mounted on the faucet. The brass adapter's fht is designed to use a typical hose washer and therefore needs no pipe dope.

12—Tighten the main assembly finger-tight. Be sure to have the outlet of the filter pointing down. If the orientation is not correct, you can sometimes add another hose washer between the Y-valve and the brass adapter to change the amount of rotation the assembly will need to seat (tighten leak-free).

Finally, attach a length of drip hose to the flushing ball-valve on the filter and turn on the faucet to check for leaks and flush the filter. If all goes well, and you haven't conjured up any of Pop Murphy's laws, you're done. The total length of the assembly in **Figure 24** is about 18 inches. If left dangling from a faucet, the assembly's weight could stress the fittings near the faucet or the fittings could easily be hit by accident and damaged. The assembly *must* be protected from damage and accidental "tweaking." In this example, two diagonal braces clamp onto the pressure regulator and stick into the soil. The braces triangulate the assembly, hold it firmly in place, and protect the Y filter from bumps and collisions. Another option is to strap the end of the assembly to a firmly secured wooden post. Mount the braces following these steps:

1—With the assembly mounted on the faucet, you can judge the angle at which the mounting bracket rods are to be pounded into the soil. The larger mounting bracket should be loosely attached to the pressure regulator. Drive one 4-foot length of 1/4-inch galvanized pipe (with one end flattened to make a point and a hole for the bolt drilled into the other end) into the ground so as to align with the large mounting bracket, then attach the rod with a bolt, two washers and a nut.

2—Next, disconnect the drip hose adapter at the end of the main assembly and slip the smaller mounting bracket over the transition nipple and onto the outlet side of the pressure regulator.

3—Pound the second rod into the soil at the proper angle. Bolt this ring mount to the rod.

The entire main assembly should be painted with an exterior or anti-rust paint. Not only will this help disguise the hardware, but the plastic will also be protected from harmful ultraviolet rays, which can cause it to become brittle and crack. Dark forest green is a common choice, but an earthy brown may be less conspicuous.

Laying Out the Drip Irrigation Hose

To get from the faucet main assembly to the flower bed, you'll need to add a length of solid drip hose. Make sure you have the correct size for in-line emitter tubing--16mm or 600 series hose with an O.D. of .67 inch.

If you have only the standard 1/2-inch solid drip hose, 18mm or 700 series with an O.D. of .708 inch, you can adapt it to in-line emitter tubing fairly easily. Purchase some compression adapters (also called

compression ring adapters), a part which is just the barbed compression ring and made to glue into standard 1/2-inch PVC fittings **(See Figure 25)**. To join the 16mm and 18mm tubings together: glue a 16mm (green) compression adapter into one end of a 1/2-inch slip-by-slip (s X s; slip means glued, not threaded) coupling and an 18mm (blue) into the other end. You can use elbows (90°), tees and couplings with any combination of compression adapters to convert from one size of tubing to the other.

Where malicious gophers lurk just below the surface, you'll want to use solid PVC pipe to get from the main assembly to the perennial bed. Gophers can hear the tiny torrent of water in a pipe, and no solid drip hose is a match for their wicked, industrious teeth. (Many people report, however, that porous pipe can withstand the onslaught of a thirsty "soil rat.")

Solid drip irrigation hose is attached to the main assembly by a fitting called a "compression X female-hose swivel." From the main assembly, the solid drip hose can curve gently down to the ground and run to the flower border. This line can be buried as deep as

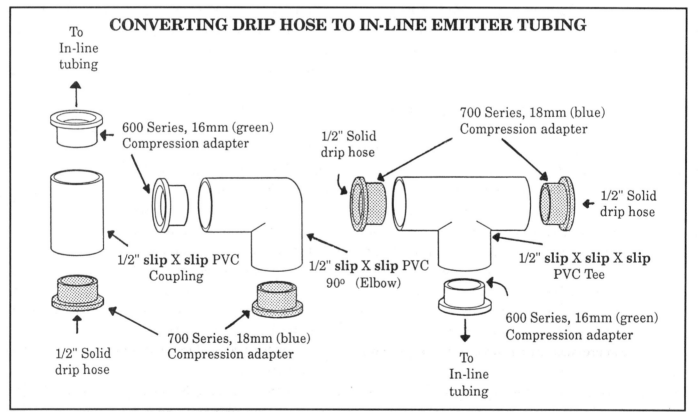

CONVERTING DRIP HOSE TO IN-LINE EMITTER TUBING

To In-line tubing

600 Series, 16mm (green) Compression adapter

1/2" slip X slip PVC Coupling

1/2" Solid drip hose

700 Series, 18mm (blue) Compression adapter

1/2" Solid drip hose

1/2" slip X slip PVC 90° (Elbow)

700 Series, 18mm (blue) Compression adapter

1/2" Solid drip hose

1/2" slip X slip X slip PVC Tee

600 Series, 16mm (green) Compression adapter

To In-line tubing

Figure 25 In-line emitter tubing isn't the exact same size as the "average" or "normal" 1/2-inch solid drip hose. When using compression fittings to join these two together, you must convert from one to the other by gluing compression adapters (also called compression ring adapters) into slip X slip PVC fittings. The in-line emitter tubing fits the green adapters (16mm), while the blue adapters (18mm) fit the 1/2-inch solid hose.

12 inches to hide it and protect it from damage.

For superior growth, I prefer to design the layout of the in-line tubing, or any drip system, to water the entire soil volume of the perennial bed. A water supply would have to be *very* limited to justify point-source watering, so, instead of being preoccupied about where the plants are located, I lay out parallel lines, or "laterals," of tubing throughout the entire bed **(See Figure 22)**. I always stock the 1/2-gph, 12-inch-interval version because I install drip systems for different clients with different types of soil. The 1/2-gph, 12-inch tubing is the only one which will water the entire length of the tubing in a sandy soil. Yet clay-loam soils can also be irrigated by turning on the system for a shorter period of time because the water easily spreads laterally. The parallel lines of tubing might be as close together as 12 inches, or closer, in sandy soils and as far apart as 20 inches in heavy clay-loams. If you have a fairly heavy soil and want to save money, buy the tubing with 20-inch or 24-inch intervals between emitters, and you'll save 25% or 40% respectively. Also, a higher flow rate, such as 1- gph, will help the water to spread wider than with the slower 1/2-gph emitters. At worst, you'll just have to run the system a little longer to get the wet spots to meet underground along the length of the tubing.

The logical time to install a drip system is during the warmer days of late spring or early summer. At these times of year, the drip irrigation hose is soft and supple and easy to work with, but it can be deceiving. Gardeners often aren't aware that drip hose shrinks considerably as the weather cools off in the fall. Because of the type of plastic used with the most common types of 1/2-inch drip irrigation hose and in-line emitter tubing, each separate length of hose can shrink by *as much as one to two feet* for every 100 feet of hose. If you lay out the drip hose during a warm day, and stretch it taut, there's a good chance, come cooler weather, that the tubing will shrink and pull out of some of the compression fittings used to make tees and 90 turns. The tendency of drip hose to shrink will stress any coupling piece, even insert and Spin Loc fittings; therefore, when laying solid drip irrigation hose in a trench (as the primary water supply from the main assembly) or on the surface with or without emitters, be

sure to snake the hose loosely along the ground or in the trench. Also, don't stake the tubing in place too securely. When inserting the landscape stakes used to position the drip hose (see the illustration on page 18 and the text on page 31) always leave a 1/8-or 1/4-inch gap above the hose to allow for shrinking and expanding. The stake will easily hold the drip hose in position without having to be hammered snugly onto the top of the tubing. At the end of each lateral, leave a little extra tubing to allow for shrinkage--up to two extra feet of tubing for every 100 feet of installed drip hose.

Many people, when they first start with drip irrigation, carefully place the emitters at the base of each plant. This is dangerous to many plants, as the upper portion of the root system near the trunk, the root's crown, is prone to crown rot, also called *Phytophthora*. Crown rot can quickly kill a plant which has been watered too much or planted in a soil with too much clay. Drip irrigation allows the gardener to avoid this problem easily by watering at least 6 to 12 inches away from the trunk. When laying out the tubing, make sure the emitters don't sit on top of an existing plant's root crown and cause rot. Also, because the roots grow well beyond the width of the plant, there is no reason to put the emitter near the trunk. Therefore, the tubing should snake slightly around the base of each plant and not be rigidly parallel **(See Figures 26 & 27)**.

All these-not-quite parallel lines join a main supply line, also called a supply "header," along the length of one side of the flower bed, usually the side closest to the faucet. For the header, I also use in-line emitter tubing so that another portion of the bed's soil gets watered. The solid drip hose from the main assembly to the edge of the flower bed is connected to an in-line emitter header by joining the two pieces with a compression coupling or Spin Loc coupler. Look at **Figure 22** to see how the header and parallel lines were placed in the Shrubberys' yard.

Each parallel line, or lateral, attaches to the header with a "compression tee." To make the header, lay out a length of in-line tubing along one side of the flower bed, cut between two emitters in the tubing where each parallel line or lateral is to be attached and wiggle the tubing into the compression holes of the

HOW TO ADD DRIP TUBING TO EXISTING PLANTS

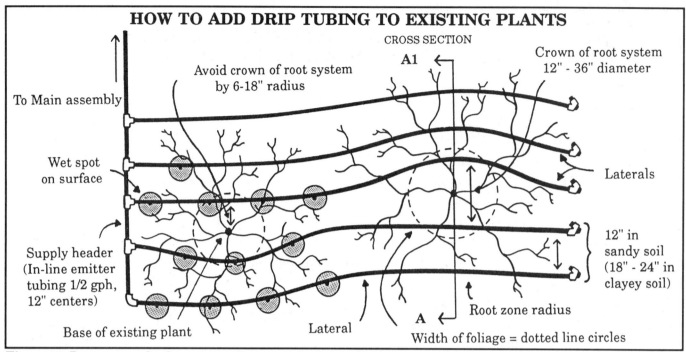

CROSS SECTION

To Main assembly

Avoid crown of root system by 6-18" radius

A1

Crown of root system 12" - 36" diameter

Wet spot on surface

Laterals

Supply header (In-line emitter tubing 1/2 gph, 12" centers)

12" in sandy soil (18" - 24" in clayey soil)

Base of existing plant

Lateral

Root zone radius

A

Width of foliage = dotted line circles

Figure 26 Be sure to snake the emitter tubing around existing plantings to avoid the crowns of the root systems, especially with drought-resistant plantings which are especially prone to crown rot.

PLACE EMITTERS AWAY FROM THE CROWN OF THE ROOT SYSTEM

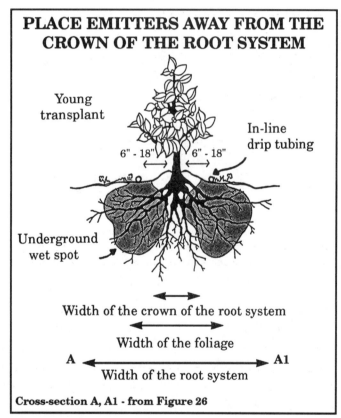

Young transplant

6" - 18"   6" - 18"

In-line drip tubing

Underground wet spot

Width of the crown of the root system

Width of the foliage

A ⟷ A1

Width of the root system

Cross-section A, A1 - from Figure 26

Figure 27 Drought-resistant plants should be planted on a slight mound or berm. Place the drip tubing slightly downhill from the top of the mound and at least 1/2 foot away from the base of the stem or trunk. Remember, the active, feeding roots are always wider than the foliage.

cross-piece of the tee. To attach the first lateral, cut halfway between two emitters on the roll of tubing and insert the tubing into the base of the first tee on the header. The second length of lateral tubing should be cut very close to an emitter and inserted into the next tee down the line in the header. (Remember to lay out the tubing loosely and leave some slack on the end of each lateral for contraction of the cold tubing.) Staggering the beginning of each subsequent lateral line like this helps to provide a more regular application of water. Instead of closing off the end of each lateral line, as in **Figure 22**, you can join the ends of each lateral with what's called a drain header, drain-down header or exhaust header. The purpose of a drain header, as I'll refer to it, is to equalize the pressure throughout the system and to speed up the flushing or draining of the system. For example, in **Figure 28**, instead of opening and flushing eight separate lines in the two sub-systems, only two end-closures, one for each sub-system, need to be opened and closed.

To plan the layout of the tubing; first, on a scale drawing of the flower bed, sketch the pattern of all the parallel laterals (the lines attached to the header) needed to irrigate the bed's soil type. With in-line emitter tubing, I usually place the parallel laterals 12

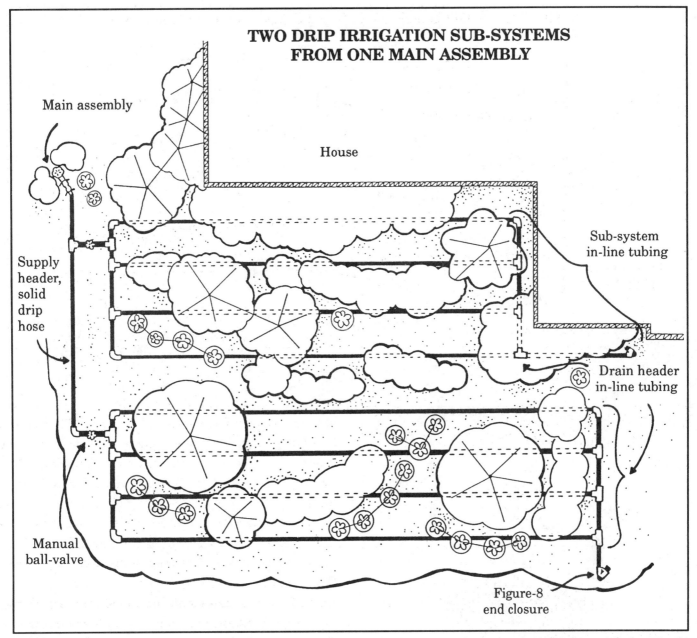

TWO DRIP IRRIGATION SUB-SYSTEMS FROM ONE MAIN ASSEMBLY

Main assembly

House

Sub-system in-line tubing

Supply header, solid drip hose

Drain header in-line tubing

Manual ball-valve

Figure-8 end closure

Figure 28 Each in-line drip irrigation system must be broken into sub-systems of no more than 326 feet, with the 1/2-gph emitters on 12-inch centers. Use ball valves to divide the tubing into sub-systems. Each sub-system can have a drain header to allow for easy flushing and draining.

inches apart in sandy soil and up to 24 inches apart in clayey soil. Next, sketch in the supply and drain headers which will supply and drain all the laterals, making sure the supply header is on the side of the bed closest to the main assembly. Remember, the recommended maximum length per valve or main assembly is 163 gph for 1/2-gph, 12-inch-interval in-line emitter tubing, or 326 feet (326 divided by 1/2 gph equals 163 gph).

If the total length of tubing in a bed is more than 326 feet, then one or more manual valves must be added to break the entire system into 326-foot, or shorter, sub-systems. To install a sub-system, add a tee for each sub-system at the appropriate spot along the header. You can insert an inexpensive barbed ball-valve made especially for drip irrigation systems to control each sub-system. Many people simply insert the in-line drip tubing over each barb on the ball-valve. (Remember, if you have difficulty inserting the drip tubing, dip the end in hot water to soften the plastic.)

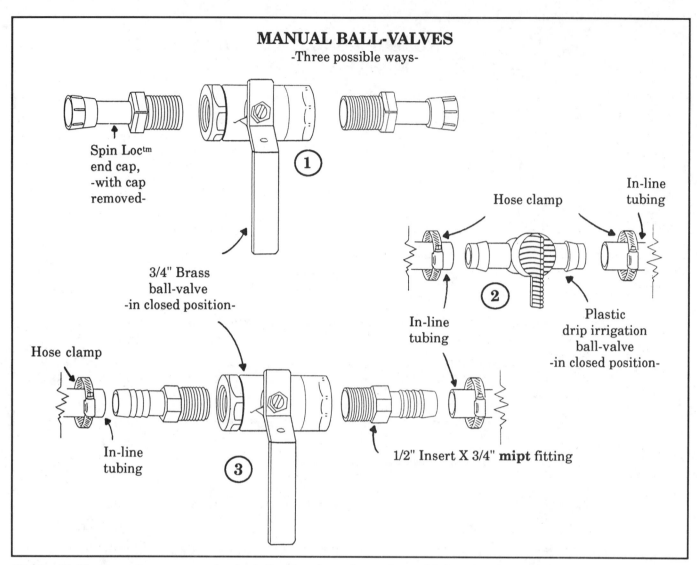

MANUAL BALL-VALVES
-Three possible ways-

Spin Loc™ end cap, -with cap removed-

3/4" Brass ball-valve -in closed position-

①

Hose clamp

In-line tubing

Hose clamp

In-line tubing

②

Plastic drip irrigation ball-valve -in closed position-

In-line tubing

③

1/2" Insert X 3/4" **mipt** fitting

Figure 29 There are three ways to plumb a ball-valve; the easiest and most likely to break or leak is the inexpensive plastic drip irrigation ball-valve.

According to the manufacturer, the barb will hold with pressures up to 60 psi. I prefer not to risk a failure while the system is on and unattended, in the tradition of Pop Murphy, and add two 60¢ hose clamps to ensure a leakproof fitting. **Figure 29** shows three ways to install a ball-valve if you have difficulty inserting the tubing onto the barbs of the plastic ball-valve. Each uses a quality ball-valve instead of the more traditional gate valve. Ball-valves seal more securely with repeated use over time than a gate valve. If your local supplier doesn't carry Spin Loc products, they most probably stock simple insert fittings (**See Figure 18**). After each shutoff ball-valve, there should be no more than 326 lineal feet of in-line tubing, but the laterals can have any configuration.

One main assembly can service two or more laterals. While the filter in the main assembly in **Figures 23 & 24** can pass 660 gph, remember, the limiting factor is the pressure regulator, with a maximum flow rate of 480 gph. In my example, 326 feet of in-line tubing passes 163 gph of water, less than one-half and nearly one-third of the pressure regulator's limit of 480 gph. This means that at any one time, the main assembly can service two or more sub-systems of in-line tubing as long as the grand total of tubing doesn't exceed 480 gph, or a total of 960 lineal feet of 1/2-gph, 12-inch in-line tubing. (Remember, each lateral or sub-system can have no more than a total of 326 feet.) When you water, close all but the number of shut-off ball-valves at the beginning of the laterals to equal

480 gph (960 feet) or less and turn on the water at the main assembly. The other laterals can be operated with a quick adjustment of the shutoff ball- valves.

Planting and Mulching for Beauty and Function

If you're adding a drip system to an existing planting, simply thread the hose beneath the foliage of each plant. To keep from clogging the hose with dirt or mulch, close the end with a figure-8 end-closure. Slip one of the circles of the figure-8 over the hose and back about 6 inches, bend over 4 inches of the end of the hose and slip the figure-8 back toward the end so the other circle catches hold of the bent-over hose. See **Figure 58** on page 62 to see how a figure-8 end-closure works. Some people make their own version of a figure-8 by simply cutting a 3-or 4-inch length of 1-inch- diameter PVC pipe. Four inches of the end of the drip hose is folded over on itself and crammed into the PVC pipe to hold it in a crimped position. Figure-8 end-closures have the habit of getting brittle and cracking over time. Many people prefer compression or Spin Loc end caps because the hose doesn't get kinked, which can lead to hose fatigue and failure, and the threaded end cap is easy to remove and rethread. I started out with figure-8s because they were cheap. I'm slowly switching over all the figure-8s to compression end caps because they perform better over the long run.

No matter how or when you install a drip system, you must flush the system at least three times during the process. Always flush the main assembly after it's attached to the faucet. Flush the open-ended hose of the main header line and then close the end. After all the lateral lines have been added to the header line, flush for several minutes and then close each lateral, starting with the line at the highest elevation or the line closest to the main assembly. Or, to simplify the whole process, plumb a drain header, with the single end-closure at the lowest end of the header. **(See Figure 28.)**

When planting a new area that's to be drip irrigated, I take the following steps:

1—Do all necessary soil preparation for the entire area.

2—Install the drip irrigation lines on top of the soil. Hold the hose in place with U-stakes (landscape staples), placed every 4 to 6 feet. Don't forget to leave the staples loose so the tubing can swell and contract throughout the year.

3—Use an overhead sprinkler to irrigate the entire area to premoisten the soil for planting. Wait one or more days for the soil to dry enough to be planted. The soil should be moist and easily crumbled, not wet and sticky.

4—Plant according to the landscape plan. Make sure each plant is 6 to 12 inches away from all emitters so as to prevent crown rot.

5—Hand-water each plant with a garden hose to settle the soil and eliminate any air pockets.

6—Mulch the planting and drip irrigation system.

Many people have a hard time with this concept when I first describe it. If they've heard or seen anything about drip irrigation, they are used to the usual system of each plant getting only one or two emitters—period. This is what I call the drought-stress approach to drip irrigation. While the plant may stay alive and even grow some, it will usually not flourish with such a small fraction of its natural root system being irrigated. The most difficult part of my approach to drip irrigation is getting people to experiment with what appears, at first, to be such a great distance between the emitters and the newly transplanted seedling, sapling or shrub. This approach depends upon the premoistening of the soil and the hand- watering of each plant before mulching. With the entire volume of the soil at an ideal moisture level, the new root hairs of the transplant could grow for days or even weeks without any additional water. By starting the drip system right away, and irrigating with small amounts of water on a daily basis, you will maintain an ideal moisture level, avoid transplant shock and promote the best growth possible. The emitters, which appear to be a "waste" because they are so far from the transplants, will soon be irrigating the fast-growing roots—remember, the roots grow in an area up to three times wider than the foliage.

No matter how you install a drip system, you should mulch. To use drip irrigation without using mulch would be like wearing pants without a belt—you get most of the effect, but without that last measure of protection. While drip irrigation doesn't waste any water through wind-blown spray or runoff, some moisture can be lost by evaporation from the soil surface. Two to 6 inches of mulch will virtually eliminate evaporation loss while at the same time hiding the tubing. Mulch also helps extend the life of the drip irrigation system by blocking harmful ultraviolet rays, which degrade the plastic hose. Be sure to keep the mulch 6 to 12 inches away from the base of the plant to avoid causing stem or crown rot.

Now you're ready to figure out how often, and for how long, to turn on your new drip irrigation system.

4 When and How Long to Irrigate

Many gardeners think it's necessary to water deeply, which often leads to the habit of infrequent but lengthy irrigations. Of course, "deeply" is a very relative word. When teaching a drip irrigation seminar, I stress maintaining good soil moisture in the top 2 feet, and I tell people not to worry about soaking any deeper. At one workshop, a woman from a local nursery pointed out that for many of their customers, watering even 6 inches down was considered radically deep. On the other hand, I've run into well-intentioned gardeners who plunge root-watering rods 18 inches deep into the soil and then run the water for hours on end, which mostly floods soils *below* the 2-foot zone. Watering should be a function of how roots really grow, not how we imagine their growth.

Deep Roots Don't Mean Deep Watering

In all the studies I've been able to find, the usual conclusion is that for the sake of quality growth, as opposed to sheer survival, the upper 1 to 2 feet of the soil accounts for over 50% of all the water a plant absorbs **(See Figures 2 & 3)**. (While many plants, including herbaceous flowers, can have roots deeper than 2 feet, these deeper roots are more important for stabilizing the plant, absorbing some micronutrients and surviving severe droughts than for supporting an abundance of growth.)

Again, here's my theory of what lies behind this interesting phenomenon. The upper layers of the soil have the highest population of air-loving soil flora **(See Figure 4)**. As mentioned earlier, the top 3 inches of the soil has nearly four and a half times more bacteria, almost eight and a half times more actinomycetes, more than twice as many fungi and five times the algae as soil 8 to 10 inches deep. The soil's flora assist in the liberation of nutrients into a soluble form which the plant can absorb. This beneficial soil life needs plenty of oxygen to fuel its activity. In addition, each consuming and liberating critter of consumption also excretes either liquid or gaseous waste. These gaseous by-products, tiny subterranean soil farts and belches, can be toxic to young roots and some of the soil's bacteria and fauna. These noxious fumes must slowly migrate up through the minute labyrinth of the soil's pore space to exhaust and dissipate into the atmosphere. The deeper the soil's flora and fauna, the longer it takes for the poisonous fumes to be "exhaled" and life-promoting gases such as oxygen and nitrogen to be "inhaled." The more aerobic layers of the upper soil horizons are quick breathers, with higher levels of oxygen and lower amounts of the noxious fumes of respiration. This allows for more rapid and extensive root growth. Deep soils aren't as aerobic; the inhalation of life-giving gases and the exhalation of toxic vapors is greatly slowed down and far less supportive of active root growth.

As mentioned earlier, plants can absorb nutrients only in a water solution. Dry soil inhibits nutrient uptake because the soil life can't thrive, and what little water is in the soil is held tightly to the soil particles. Drying out the soil between irrigations means that nutrients also "dry up." When water is supplied to the soil to the point of saturation, the roots and air-loving soil life may be stressed or killed from waterlogging, and nutrient uptake is reduced due to the lack of air. As the wet soil dries, it takes some time for the aerobic flora to repopulate the soil. Either extreme, too wet or too dry, causes a biological lag before the roots get their best meals. Infrequent and deep irrigations produce two periods in the cycle in which the plant's growth is reduced or hindered—during the drought stage and when the soil is saturated.

Soils Aren't Always Deep

Remember, deep glacially deposited topsoil is more the exception than the rule. Even if you have such a deep, loamy soil, remember that the majority of moisture and nutrient absorption for quality growth still happens in the top 2 feet of the soil. More typically, most suburban yards have a very shallow layer of topsoil. If, for example, there is a continuous deposit of rich orange clay just 12 inches under the ground, then the 12-inch layer of topsoil is the only place where your plant's roots will be absorbing water and nutrients.

Irrigate Frequently

As mentioned earlier, frequent waterings

produce the the best-looking foliage growth, the most abundant bloom and the highest yields.

I prefer frequent or daily watering to replace the moisture lost due to evaporation from the soil and transpiration from the plant's leaves (called evapotranspiration, or ET), plus an amount that represents enough extra water for gorgeous ornamental growth. Each spring or early summer, I don't start watering until the soil reaches an ideal moisture level: not too anaerobically wet and not too dry, barely as moist as a wrung-out sponge. The frequent irrigations are meant to maintain this ideal moisture level and eliminate the stress of soil which is too wet or too dry. In areas with summer rain, you should wait until the soil naturally drains to an ideal moisture content and then water frequently with small amounts of water until the next rain saturates the ground.

Frequent Irrigation Can Use Less Water

Daily irrigation doesn't mean using countless gallons of extra water. In fact, with infrequent irrigation, it takes a certain amount of water just to rehydrate the soil before the plant can even make use of the moisture. Oddly enough, infrequent waterings can use *more* water than the same planting would receive with frequent, even daily, irrigation.

Years ago, for example, I planted a drought-resistant landscape, with plants such as lavender, santolina, rockroses, and rosemary, for my neighbor Mary Burke. I followed the same steps for soil preparation and preirrigation outlined on page 31. The day after planting, the timer was set to irrigate each zone for 15 minutes. After the risk of transplant shock was over, the irrigation lines were turned on each day for only 8 minutes. The plants flourished, even though each 1/2-gph emitter was distributing the paltry amount of 7 *tablespoons* of water per emitter each day. Contrast this with a nearby garden with a similar soil and lavender plants arbitrarily watered only twice a month for four hours. This amounts to 2 gallons per emitter for the two-week period, or just more than 18 tablespoons of water per day—more than twice the water used in Mary's flourishing landscape.

No matter how you use drip irrigation, frequently or every once in a while, it will always be more efficient than any sprinkler you're currently using. All sprinklers, except the most modern of micro- or mini-sprinklers, apply water faster than many silty and clayey soils can absorb it. This leads to anaerobic puddling and runoff, especially on steep slopes. *Every* sprinkler is vulnerable to wind- and sun-induced losses, with as much as 25% of the water wasted. In general, sprinklers are rated at an overall efficiency of 75% to 80%, compared with drip irrigation's 90%. (Furrow irrigation can have an efficiency rating as low as 50%.)

How Long to Water

There are two general approaches to watering, the empirical and the more analytical, which uses the evapotranspiration rate as a guideline. Each works, but the ET-rate-based approach can be far more accurate and water conserving.

The most immediate, or empirical, way to understand your soil's response to drip irrigation, and to determine how long to leave the system on, involves digging. Even after doing your experiment with the milk jug, you should test for the drip system's underground pattern of moisture. Turn on the drip system for an hour, then turn off the hose and dig a number of small holes in the flower bed to see how deep and to what width the water has soaked in. Then turn the system on for another hour, to equal a test total of two hours, and check to see how much farther the water moves. Do this for several more intervals of time and observe vague changes in the wet spot. This test will reveal the shortest length of irrigation time to produce the widest wet spot, based on which test hole revealed the widest spread of the wet spot during the elapsed time. Without doing this test, you'll just be guessing in the dark.

ET-Based Irrigation

Another approach involves using the ET figures for your local climate. The ET rate combines the amount of moisture lost from the soil's surface (evapo-) and the foliage (-transpiration.) It represents the amount of water a plant uses, regardless of the method of application, and is determined with a formula which factors the temperature, wind speed, humidity and percentage of ground covered by foliage. The rate varies considerably, depending upon the season—the coolest days of early spring, windy days in March, hot, muggy summer weather or dry winds in August. Your local Cooperative Extension office should be able to tell you either the current week's ET rate or the month's average rate; both are expressed in inches per day or month.

The chart in **Figure 30** shows the daily water use for ten different ET rates. Remember, the amount of water needed to replace the ET losses depends upon the amount of soil covered (like a shadow) by the planting's foliage. If the plants are young, the ET rate is less, corresponding to the smaller area of coverage. With a mature flower border, the coverage is complete and all you need to determine is the total square footage of the border. For example, a 5- by 20-foot border (100 square feet) uses 18.7 gallons of water per day during a hot day

when the ET rate is equivalent to 9 inches of water per month (derived by cross-referencing the columns in **Figure 30**). If you still prefer to water once a week, multiply the daily ET rate by seven to determine the total amount for the weekly watering.

To determine the length of each day's watering, take the total amount of water the flower border requires and divide by the total flow of the drip irrigation system. Consider again the illustration of Mr. and Mrs. Shrubbery's 5- by 20-foot border with a daily ET rate of 18.7 gallons. **(See Figure 22.)** Since the total length of in-line emitter tubing in the bed is 84 feet (this figure is reached by adding together one header 4 feet long and four 20-foot laterals), the sum flow of the system is 52 gph (84 1/2-gph emitters times their actual flow of .62 gph). (See the **Appendix**, the Maximum Length of Tubing Chart.) Thus, dividing the

Daily Water Use (In Gallons per Day)
BASED ON VARIOUS EVAPOTRANSPIRATION RATES

Square Feet of Plant Cover	ET Rate (in inches/month)									
	1"	2"	3"	4"	5"	6'	7'	8"	9"	10'
1 sq. ft.	0.0187	0.0374	0.062	0.083	0.104	0.125	0.145	0.166	0.187	0.208
4 sq. ft.	0.075	0.15	0.248	0.332	0.416	0.5	0.58	0.664	0.75	0.832
10 sq. ft.	0.187	0.374	0.62	0.83	1.04	1.25	1.45	1.66	1.87	2.08
75 sq. ft.	1.403	2.805	4.65	6.225	7.8	9.4	10.875	12.45	14.	15.6
100 sq. ft.	1.87	3.74	6.2	8.3	10.4	12.5	14.5	16.6	18.7	20.8
200 sq. ft.	3.74	7.480	12.4	16.6	20.8	25.	29.	33.2	37 4	41.6
300 sq.ft.	5.61	11.22	18.6	24.9	32.2	37.5	43.5	49.8	56.1	62.4
1 acre solid cover	815	1629	2701	3615	4530	5445	6316	7231	8146	9060

Figure 30 A number of daily ET (evapotranspiration) rates. Ask your local Cooperative Extension agent for the ET rate for your growing season. These rates are just a starting point, you can adjust the amount up or down depending upon your water supply and how much growth you desire. The square footage of cover represents the surface of the soil covered by the shadow (at noon) or "foot print" of the plant's foliage, regardless of the shape or type of the foliage.

daily water need of 18.7 gallons by 52 gph yields .36 hour, or 22 minutes per day (.36 of an hour times 60 minutes.) If you want to water once per week, then multiply 22 minutes per day times seven days to get a weekly watering time of 154 minutes, or nearly three hours.

The above calculation is fairly simple because I chose the example of emitters on 12-inch intervals and the rows of tubing 12 inches apart. What if your system has emitters or tubing at odd intervals? The calculation in **Figure 31** may appear difficult at first, but it is rather straightforward. Another advantage of this calculation is that the rate of irrigation is converted to inches, which has been a more common way to gauge irrigation rates. The first part of this formula is another way to calculate the rate of application in inches per hour.

Rate of Application Formula

$$RA \text{ (Rate of Application)} = \frac{231.1 \times GPH}{\text{sq. in. spacing}}$$

Figure 31 The rate of application (RA) formula is used to calculate the total flow, per square foot, of emitters spaced at odd intervals such as 18 inches apart along the drip tubing 24 inches apart on the ground.

For example, if you have 1-gph emitters on 18-inch centers on the tubing and the lines of tubing are 24 inches apart, multiply 18 times 24 to get 432 square inches. Divide 432 into the product of 231.1 times .92 (the actual flow rate of the 1-gph emitters.) **(See Appendix.)** This gives a result of .49 inch per hour, the average irrigation rate in inches per hour for the entire system.

Many gardeners are used to irrigation guidelines given in inches per week. With the knowledge of your system's inch-per-hour rating, it's easy to figure how long to leave the system on to satisfy the plants. If, for example, the local farm advisor recommends 2 inches per week for peach trees, then the above system would have to run slightly more than four hours per week (2 inches divided by .49 inches per hour = 4.08 hours). This four hours of irrigation could be all at once, or two irrigations per week of two hours each, or even, although I don't recommend it, 16 hours of irrigation once a month.

If you want to irrigate on a gallonage basis, the results of the RA formula must be converted to gallons per hour per square foot (g/hr/sq. ft.) To convert, divide the RA by 12 to get the cubic foot (cu. ft.) amount of water per hour (cu. ft./hr). In this case, .49 divided by 12 equals .04 cu. ft./hr. To convert to g/hr/sq. ft., multiply .04 cu. ft./hr times 7.48 (the number of gallons per cubic foot.) For this example, .04 times 7.48 equals .3 g/hr/sq. ft. (Take a deep breath.)

So, returning to the example of the 100-square-foot border which requires at least 18.7 gallons per day. If the tubing is configured like the **Figure 31** example with 1-gph emitters (18 inches in the tubing by 24 inches between the lines) it could apply 30 gph (.3 g/hr/sq. ft. times 100 sq. ft.) The length of irrigation equals .62 hour (18.7 total gallons divided by 30 gph) of irrigation per day, or 37 minutes **(see Figure 32)**.

I hope this math doesn't make your head hurt as much as it did mine initially! The first time may seem like a mind-bender, but if you work them through a few times, these calculations will soon become routine. Besides, once you've struggled through the initial math, you will have established a basic guideline for all of your subsequent irrigation. From that point on, the length of irrigation is adjusted up or down depending upon the quality of growth desired or on your water supply. If you want to fine-tune the system throughout the growing season, once you know the flow rate for each system, you can just do the math for the ET rate for different seasons.

Refining ET-Based Irrigation Rates

The above ballpark approaches to ET-based irrigation are the simplest, but the formulas ignore two important realities: all plants are not identical in their ET rate, and drip irrigation, in spite of its high efficiency, isn't perfect. There is yet another formula which makes ET-based irrigation even more accurate and less of an assumption. This formula for computing daily water needs is outlined in **Figure 33**.

The .623 figure in the formula is a constant which mathematically adjusts the varying square footage of root zones to the ET rate, and functions to convert the result to gallons. This number was derived by dividing 12 (inches) into the number of gallons in a

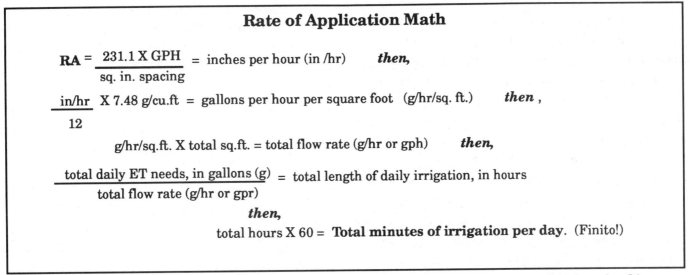

Rate of Application Math

$$RA = \frac{231.1 \text{ X GPH}}{\text{sq. in. spacing}} = \text{inches per hour (in /hr)} \qquad \textbf{\textit{then,}}$$

$$\frac{\text{in/hr}}{12} \text{ X 7.48 g/cu.ft} = \text{gallons per hour per square foot \ (g/hr/sq. ft.)} \qquad \textbf{\textit{then,}}$$

$$\text{g/hr/sq.ft. X total sq.ft.} = \text{total flow rate (g/hr or gph)} \qquad \textbf{\textit{then,}}$$

$$\frac{\text{total daily ET needs, in gallons (g)}}{\text{total flow rate (g/hr or gpr)}} = \text{total length of daily irrigation, in hours}$$

$$\textbf{\textit{then,}}$$

$$\text{total hours X 60} = \textbf{Total minutes of irrigation per day.} \ \text{(Finito!)}$$

Figure 32 This is a specific example of the math involved in computing the rate of application (RA) formula. It's not as scary as it looks at first.

Daily Irrigation Need Formula

$$\text{DN (Daily Needs), gallons} = \frac{.623 \text{ X \textbf{Area of the root zone} X \textbf{Plant Factor} X \textbf{ET rate}}}{\text{Efficiency of drip irrigation by climate}}$$

Figure 33 The daily needs formula (DN) is a more specific way of calculating the irrigation needs of your plants, and is more accurate than just using the table of ET rates on page 35. You needn't use this formula if it scares you; just use the simpler ET table and adjust the length of irrigation based upon your experience.

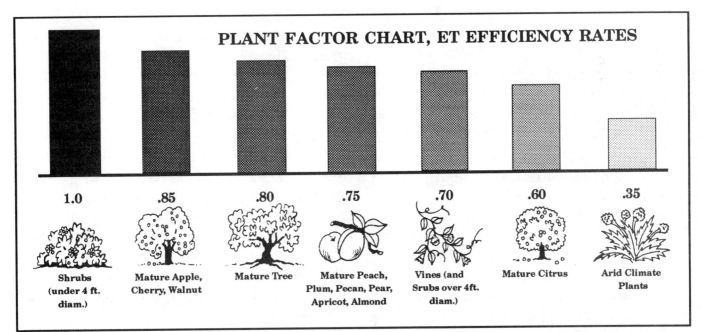

Figure 34 Not all plants have truly identical ET rates. Some plants are more moisture-conserving than others. This chart lists the ET-rate efficiency of a number of plants and trees. These rates of efficiency are still somewhat imprecise, but offer more accuracy than the rates listed in the ET chart on page 35.

cubic foot of water (7.48 gallons).

Since my approach to drip irrigation involves watering the entire root zone, the area of the root zone is equal to the area covered by the tubing plus the lateral movement of water away from the outermost emitters. In most cases, this is the area of the entire bed, container or planting. Don't forget that the true width of the root zone is one to three times wider than that of the foliage. If you're watering an individual shade tree which is isolated from other plantings, then simply calculate the area of the root zone. For circles, the area equals the radius time the radius (that is, the radius squared) times 3.14. For some, it's simpler to square the diameter and multiply by .7854.

Plants actually transpire at different rates; depending upon their size, type and age, and each plant can have a different factor of ET efficiency. For the sake of this calculation, shrubs smaller than 4 feet, unless an ET factor is known, are rated at 100%, or a 1.0 factor. Generally speaking, drought-resistant plants have a .35 plant factor. This means they use 65% less water than a normal water-loving plant. Refer to **Figure 34** for more plant factors.

Finally, the ET rate is based on **Figure 30**. Instead of using the daily gallon rate from the 1-square-foot column, use the daily ET rate in inches. This can be determined by dividing the ET (in inches per month) figure by 30 days. For example, in **Figure 30,** 1, 2, 3,4, 5,6, 7, 8, 9 and 10 inches per month become .03, .07, .1, .13, .17, .2, .23, .27, .3 and .33 inch per day, respectively. Choose an ET rate on the high end of your monthly ET.

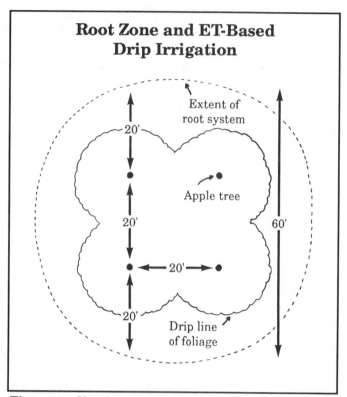

Root Zone and ET-Based Drip Irrigation

Figure 36 Use this sample planting plan as the basis for calculating daily needs (DN) in the example given in Figure 37.

All of this is to be divided by the efficiency of the drip irrigation system. While I mentioned earlier in the book that drip systems are 90% efficient, this is really just an averaged figure. The efficiency of any irrigation system is affected by the climate. **Figure 35** shows the efficiency rates for three simplified climate zones. When in doubt, select an efficiency rate from a harsher climate than yours.

For this example, let's consider the irrigation of four apple trees planted in a moderate climate in a square pattern on 20-foot centers (**see Figure 36**). The root zone and the drip system cover a total of 3600 square feet (60' X 60'). From **Figure 34,** the plant factor is .85; the ET rate is .2 (6 inches per month) and the efficiency of the drip system is .9. See **Figure 37** for the calculation.

Efficiency of Drip Irrigation Systems

Climate	Factor
Dry-Warm/Hot	.85
Moderate	.90
Humid	.95

Figure 35 While drip irrigation is the most efficient way to irrigate, it's efficiency is still influenced by climate. A factor of 1.0 would indicate an absolutely efficient system.

Daily Irrigation Needs

$$\frac{.623 \text{ X } 3600 \text{ X } .85 \text{ X } .2}{.9} = \begin{array}{l} 424 \text{ gallons per day} \\ \text{for all four trees} \end{array}$$

Figure 37 This illustrates the math for calculating the daily needs (DN) outlined in the text.

Now you've got the actual water needs of the trees. Next, total the number of emitters in the system and calculate the flow-rate per hour, then divide into 424 gallons per day to get the total number of hours the system must run each day.

ET Facts for Trivia Buffs

Even this last formula for using ET rates to calculate irrigation rates relies upon generalized assumptions about plants. Yet plants, like people, are unique; each has a different physiology and distinctive needs.

David Johnson, while working as a technician at the Laboratory of Climatology at Arizona State University, developed, after considerable library research, some very interesting figures about maximum ET rates for various crops and plants. Who would suspect that the more petite pomegranate (*Punica granatum*), with a mature height of less than 12 feet, would transpire more water (35 inches per year) than a lofty Aleppo pine tree (*Pinus halepensis*), with a mature height of 30 to 60 feet and a maximum transpiration rate of 25 inches per year.

I was surprised to find from Johnson's data that an olive tree's annual transpiration of 25 inches is higher than that of a pineapple (20 inches). Perhaps this difference is due to the thick, waxy covering on the pineapple's leaves, or the humidity of a tropical climate. Or it may simply be due to the large difference in total surface area of leaves of olive trees and pineapple plants. While the exact mechanism is uncertain, the discrepancy is striking.

More remarkably, a number of Sonoran desert scrub trees, including several species of *Acacia*, growing in one of the drier climates in the country have a maximum transpiration of 35 inches, more than the average of 33 inches for deciduous fruit trees. This may be due to the capacity of many desert plants to grow very extensive and lengthy root systems. While some desert plants use more moisture than some humid-climate plants, they compensate by having root systems that are better able to explore for and secure elusive soil moisture.

In other cases, desert plants behave the way we would expect. Desert cacti transpire no more than 15 inches of moisture per year and need less than 1 inch of water to survive, compared with an orange tree's maximum use of 35 inches and survival need of 11 inches of water.

For more of Johnson's fascinating information, see the **Appendix**.

5 Hiding and Expanding Your Drip System

The main assembly for the flower border or any other section of the landscape can be discreetly hidden at or below the soil's surface (**See Figure 38**). This makes for a less cluttered-looking garden, but beware. You may easily forget to flush the filter or bring the assembly indoors for the winter (if necessary in your area). Also, please note, when considering a "buried" assembly: if the main assembly is below ground level, all plants to be irrigated must be downhill from the level of the assembly (at least 12 inches lower) in order for the atmospheric vacuum breaker and check-valve to be effective. If this is not possible in your yard, you can add a brass vacuum breaker (see page 46) back near the house where the main irrigation line joins the house's water pipes. Be sure the anti-siphon device at the house is 12 inches or more above the highest part of the irrigated landscape. Instead of a faucet, a ball-valve is plumbed into this assembly to turn the system on and off.

Due to the space limitations of this assembly housing, you won't be able to rotate the filter easily to thread it onto the assembly. A union (**See Figure 40**) allows you to add the main assembly just after the ball-valve without having to rotate any part. The union has three parts: one end with fipt threads and nonpipe male threads with a flat seat, another end with fipt threads and a flat seat with an O-ring gasket, and a large ring with nonpipe female threads which joins the flat faces of the two other parts together into one leak-free fitting.

Before you begin, gather your wits about you, along with some plumbing tools and all the parts you'll need. In **Figure 39** I've provided a shopping list for

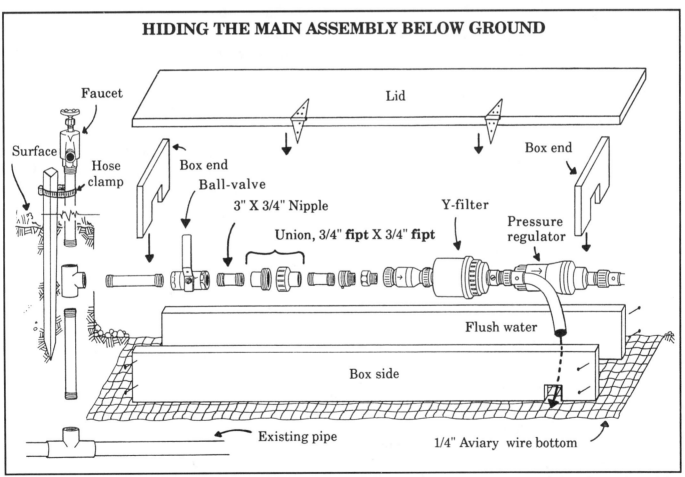

HIDING THE MAIN ASSEMBLY BELOW GROUND

Figure 38 Some gardeners prefer to have their drip irrigation main assembly hidden in a box just below the surface of the soil. Remember the "out-of-sight, out-of-mind" syndrome, which may cause you to forget to flush the filter or bring the main assembly indoors during cold winters. If you skip the protective wire mesh bottom, gophers will fill the box in less than one night with their "throws."

both the tools and the plumbing parts you'll need. Be sure to check your purchases twice so that universal law of plumbing, PMLP Numero Uno, doesn't rear its ugly head.

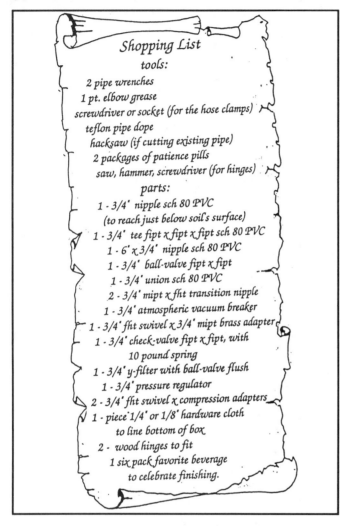

Shopping List

tools:
2 pipe wrenches
1 pt. elbow grease
screwdriver or socket (for the hose clamps)
teflon pipe dope
hacksaw (if cutting existing pipe)
2 packages of patience pills
saw, hammer, screwdriver (for hinges)

parts:
1 - 3/4" nipple sch 80 PVC
(to reach just below soil's surface)
1 - 3/4" tee fipt x fipt x fipt sch 80 PVC
1 - 6" x 3/4" nipple sch 80 PVC
1 - 3/4" ball-valve fipt x fipt
1 - 3/4" union sch 80 PVC
2 - 3/4" mipt x fht transition nipple
1 - 3/4" atmospheric vacuum breaker
1 - 3/4" fht swivel x 3/4" mipt brass adapter
1 - 3/4" check-valve fipt x fipt, with
10 pound spring
1 - 3/4" y-filter with ball-valve flush
1 - 3/4" pressure regulator
2 - 3/4" fht swivel x compression adapters
1 - piece 1/4" or 1/8" hardware cloth
to line bottom of box
2 - wood hinges to fit
1 six pack favorite beverage
to celebrate finishing.

Figure 39 Here's your shopping list for the project to construct a main assembly which is cleverly hidden below the surface of the garden.

Now You Can Begin

1—Pull the soil away from the base of your existing faucet riser. Carefully dig down to where you find the riser (actually a very long nipple) attached to a threaded fitting. This might be a threaded "tee" if the riser is in the middle of a long irrigation line. Or there might be a threaded elbow (a 90° fitting or "ell") if it's the end of an irrigation line. Be sure to dig *below* the level of the fitting to prevent soil from spilling into the open pipe when the riser is removed.

2—All risers should be attached to a post to keep them from "tweaking" the underground fitting. Loosen the fastening so you can unthread the riser. (If there is no post, you'll need to add one after you've finished re-plumbing the riser.)

3—With a pipe wrench, grasp the riser above the soil level and turn it counterclockwise to remove it from its fitting on the underground pipe. When backing the riser out, use another wrench (watch out for PMLP #3) to hold the fitting, to make sure it's not tweaked. If the fitting appears rusty, treat the base of the riser with some Liquid Wrench™ or WD-40™ to loosen the threads.

If the riser is just a white plastic pipe that is glued, not screwed, into a white fitting, you should cut the irrigation pipe and glue a new schedule 80 threaded fitting (a gray-colored fitting) into the existing irrigation pipe. Schedule 80 PVC pipe fittings, while not as sturdy as metal pipe fittings, are many times stronger than the white schedule 40 fittings. Be sure to use the special glue made to bond the gray schedule 80 fittings to white PVC pipe; check the description on the label or ask the hardware store clerk for assistance. Try to avoid manifesting PMLP #7. Make sure the fitting you glue has fipt threads on the outlet portion.

4—Now, using either threaded schedule 80 parts or galvanized metal fittings, add a new, shorter riser nipple to reach 3 inches below the natural level of the soil. (Be sure to use Teflon™ pipe dope, in spite of PMLP #8, to seal all the threads mentioned in these instructions.) The length of this nipple will vary according to how deep the original irrigation pipe was laid. Install the nipple with two pipe wrenches, one grasping the nipple between the threads, the other firmly holding the fitting below ground to prevent it from getting tweaked. (If any dirt fell into the old irrigation pipe before the new, short riser was added, turn on the water briefly before rethreading the old riser to help flush the dirt out.)

5—Install a threaded 3/4-inch tee (fipt X fipt X fipt) to the bottom of the old riser with the faucet. Install the tee so that the pass-through openings will be straight up and down and the "leg" will be parallel to the ground. As the tee gets snug, make sure the "leg," when you're looking down on the faucet, is perpendicular to the direction of the faucet.

6—Thread the old riser on top of the new fipt tee. (Don't drop it, or PMLP #5 will come into play.)

7—Next, add a 6 inch nipple to the "leg" of the fipt tee.

Be sure to hold the riser securely while threading on the nipple.

8—Attach one of the fipt openings of the ball-valve to the end of the nipple. You will have to dig a hole deep and wide enough to rotate the ball-valve onto the nipple.

9—Add a 3-inch nipple to the end of the union with the fipt threads and the seat with the O-ring gasket. Be sure to slip the union's female-threaded ring over the end of the union before adding the nipple. Keep an eye on the O-ring, as it can easily fall out during the precarious act of plumbing.

10—Because the simplest brass atmospheric vacuum breaker has hose-threads, you'll have to thread an adapter into the outlet side of the union and onto the inlet side of the check-valve. The adapter looks much like a short nipple but has, on one end, male pipe-threads to screw into the union and, on the other side, male hose-threads. Use some Teflon pipe dope only on the mipt portion. Hand-tighten the atmospheric vacuum breaker onto the adapter's mht end. Add the next adapter, a 3/4-inch brass fht X mipt fitting, to the vacuum breaker. Hand-tighten.

11—Apply some pipe dope to the brass adapter's mipt end and carefully twirl on a complete main assembly as described in the chapter titled Your First Drip Irrigation Project. Be sure not to misalign the assembly and ruin the threads by "cross-threading" the plastic threads on the check-valve.

12—Tie or fasten the riser back to the support post. If there was no supporting post, carefully hammer in a wooden 2" X 2" (with a pointed end for easier insertion), a length of 3/4-inch rebar, or a length of 1/2-inch galvanized pipe with one end pounded flat to make it easier to drive into the ground. Using large hose clamps, strap the riser to the post in two places—one near the faucet and the other on the lower half of the riser.

13—Join the finished main assembly to the portion of the union that's attached to the ball-valve. Double-check to make sure the O-ring didn't fall out. Also make sure both flat surfaces of the union are clean and free of any dirt or mulch. As you tighten the union's female-threaded ring, make sure the filter's discharge valve is parallel to the ground.

14—Add a swivel X compression fitting to the mht fitting of the ball-valve at the end of the filter. Insert a length of solid drip hose into the compression fitting. This allows you to flush the filter without filling up the box with water. Place the end of the flush hose near a plant to utilize the water. Stop, relax and reward yourself.

Now construct a box to surround the main assembly. The ball-valve remains outside the box to make it easy for you to reach the valve. Be sure to cut a U-shaped slot in the box to fit over the pipe between the ball-valve and the union, over the drip hose as it exits toward the flower bed and over the filter's discharge hose. All worries of PMLP #11 aside, if you have moles or gophers, before you place the box around the assembly, lay a doubled-over piece of 1/4-inch metal hardware cloth on the ground. The hardware cloth helps keep the box from filling up with soil from these tunneling pests. Or pour a thin slurry of concrete into the bottom of the box.

The union allows you to remove the main assembly for repairs or winter storage without digging up the box. To remove the main assembly, close the ball-valve, unscrew the swivel which attaches the drip irrigation hose to the filter, unscrew the swivel fitting from the filter's flush discharge and unthread the large outer ring of the union—you can then lift the main

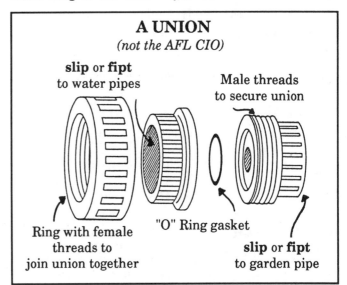

Figure 40 The union allows you to take a main assembly off the main water supply line, for repairs or storage in the winter, without having to take the entire assembly apart. The large outer ring joins the two sections together, and the gasket seals against leaks.

A GROUND LEVEL MAIN ASSEMBLY

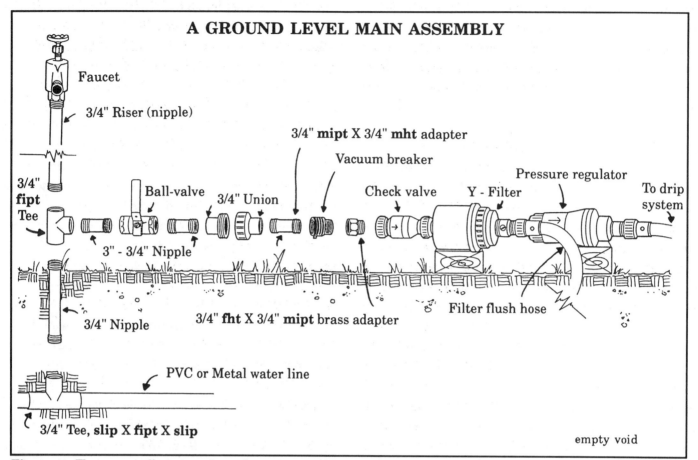

Figure 41 These parts allow you to have the main assembly near the ground, where it won't show from a distance but will be more accessible for cleaning the filter. Be sure to place the blocks of wood under the filter and pressure regulator to reduce the stress on the fittings.

assembly out of the top of the box.

With the assembly hidden, you will probably need some kind of reminder system to help you remember to flush the filter periodically and bring the main assembly indoors in cold-winter climates.

Plumbing the Main Assembly at Ground Level

To give your main assembly a low profile while avoiding the "out-of-sight, out-of-mind" drawback of the hidden main assembly, you can use the same configuration of parts mentioned above but plumb them to lie nearly on the soil's surface, at the base of the faucet. A little ruffle of evergreen herbs or shrubbery will veil the assembly from view in the garden but still leave it visible enough to be a reminder for seasonal maintenance. If you're the kind of person who requires visual reminders to get things done, this is a still-visible but more elegant solution than hanging the main assembly off the faucet, 18 inches above ground.

Figure 41 shows that this approach is very similar to plumbing the main assembly below ground. Follow steps #1 through #14 as mentioned above. The exception, in step #4, is to add a longer nipple to bring the tee, and subsequently the main assembly, 2 to 3 inches *above* the ground. The nipple in step #7 can also be reduced to just 3 inches, since you're not building a box around the assembly. You can skip the addition of a union (steps #8, 9 and 10) if you want, but you'll have to dig a shallow hole to allow for the rotation of the Y-filter while threading it onto the check-valve. The union, however, makes it much simpler to remove the main assembly in cold-winter climates. Be sure to add, as illustrated, some blocks of wood beneath the assembly to keep it from tweaking any of the parts leading back to the faucet's riser. You must paint the entire assembly with anti-rust exterior metal paint to help delay the brittling of the plastic by the sun's ultraviolet rays. Beneath the entire assembly, put down a weed barrier—a piece of carpet with the less noticeable brown backing facing up, ten sheets of newspaper or one thickness of corrugated cardboard covered with an attractive mulch, or a piece of woven landscape fabric, which allows water and air but not weeds to pass through.

Attaching the Main Assembly to the Hose

Perhaps the easiest approach to the main-assembly is to attach the main assembly permanently to the end of a length of garden hose. The other end of the hose can be attached to one side of a Y-valve at a faucet and the main assembly can then be taken to whichever section of the landscape needs watering. In this case, the main assembly is constructed exactly as it would be for mounting on a faucet, except the fht fitting at the beginning of the assembly is threaded onto a metal in-line ball-valve with hose-threads and then onto the mht end of the hose **(See Figure 42)**. The ball-valve permits you to turn on the water at the main faucet, but not waste any water before you've joined the assembly to one of the drip systems in the landscape. It's imperative that you *do not* pick up the assembly by grabbing the end of the hose. The weight of the

assembly will quickly crack the hosing. Lift the assembly up by the filter, which is the easiest part to hold and is more or less in the middle. Again, be sure to apply a protective layer of paint to the assembly.

Out in the landscape, each separate drip system begins with a fht swivel X compression fitting. These fittings are commonly sold with a combination gasket and cone-shaped screen in the swivel. The screen must be left in the swivel to keep bugs, worms and other disgusting creatures and dirt from entering the drip hosing when the main assembly is not attached.

You can streamline your watering by adding brass speed-couplings to both the end of the main assembly and all fht swivels in the landscape. Speed-couplings work much the same way the gizmo you use to inflate your car's tires attaches to the end of the air compressor's line. I always recommend solid brass speed-couplings over the plastic versions, even the European versions, because they can take much more cold and abuse without breaking. To add a speed-coupling to the end of the main assembly, add a mipt X mht transition fitting **(See Figure 42)**, then thread the female threads of the brass speed-coupling's connector onto the male hose-threads of the transition fitting. At

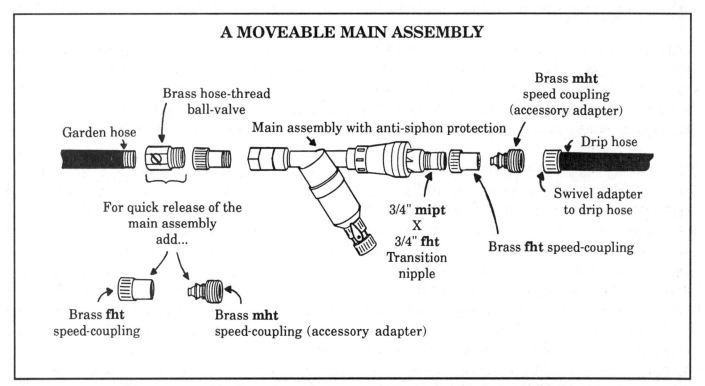

Figure 42 A moveable main assembly allows the gardener to use one main assembly for numerous drip irrigation systems or subsystems. Because the parts for a main assembly are rather expensive, the moveable assembly can save you plenty of money. Make sure the drip lines are protected from bugs when not attached to the main assembly.

the beginning of each drip system, add a permanent male hose-threaded speed-coupling, sometimes called an accessory adapter, into the fht swivel. In order to attach the male hose-threaded piece you will have to reverse the typical position of the cone-shaped screen. Face the *convex* side of the screen into the throat of the fht swivel; otherwise, the cone will block the tightening of the male speed-coupling. Then, you can attach or unfasten the main assembly with a quick snap or release of the speed coupling. The small cost of the speed couplings will be repaid many times over by their convenience.

Using One Main Assembly for Several Sections of Your Landscape

The best way to save money is to design your drip system (and, if possible, your landscape plantings) to allow for one main assembly to service several or all drip irrigation systems. This is usually most easily done at a faucet or water pipe next to the house. The idea is to have one main assembly serving several rigid plastic PVC pipes, which are buried in trenches leading out to each separate drip system.

If this centralized main assembly is 12 or more inches lower than any portion of all the separate drip systems, you can't use a single check-valve. Even a brass atmospheric vacuum breaker **(See Figure 43)** has to be 6 inches higher than any portion of the drip system. The only safe anti-siphon device for this setting is a brass backflow preventer **(See Figure 44)** with a double check-valve. These are expensive ($125 for the 3/4-inch model and nearly $140 for the 1 inch size) but worth the price compared with the alternative of contaminating your family's drinking supply. These seemingly expensive anti-siphon devices can be cheaper than installing a lot of brass atmospheric vacuum breakers with manual valves at various points in the landscape. For example, if your landscape needs six or more vacuum breakers, at nearly $21 each, brass atmospheric vacuum breakers with manual valves are more costly than a single $125 backflow preventer.

For this centralized main assembly, you'll need to know if you plan to irrigate any two sections at once. If so, you'll have to install the larger 1-inch versions of the anti-siphon device, filter and all connecting parts to

maintain sufficient flow. Remember, a 3/4-inch filter can pass only only 660 gallons per hour, compared with 1200 gph for a 1-inch filter. In most typical yards, the pressure regulator is best installed at the end of the PVC pipe and the beginning of the drip hose to reduce friction loss problems.

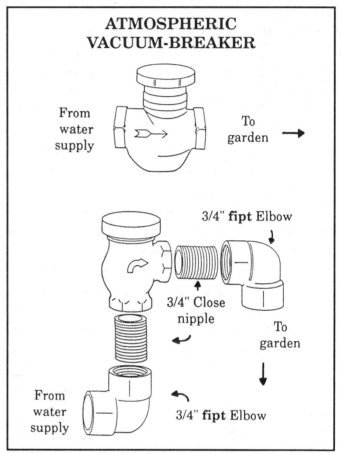

Figure 43 Two versions of brass anti-siphon valves, which protect the house's water supply much as a check valve does. These must be 6 inches higher than the entire irrigation line.

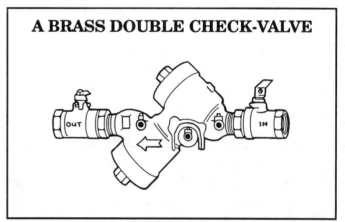

Figure 44 The highest-quality, and most expensive, backflow preventer—a double check-valve. This can be placed level with or lower than the irrigation line.

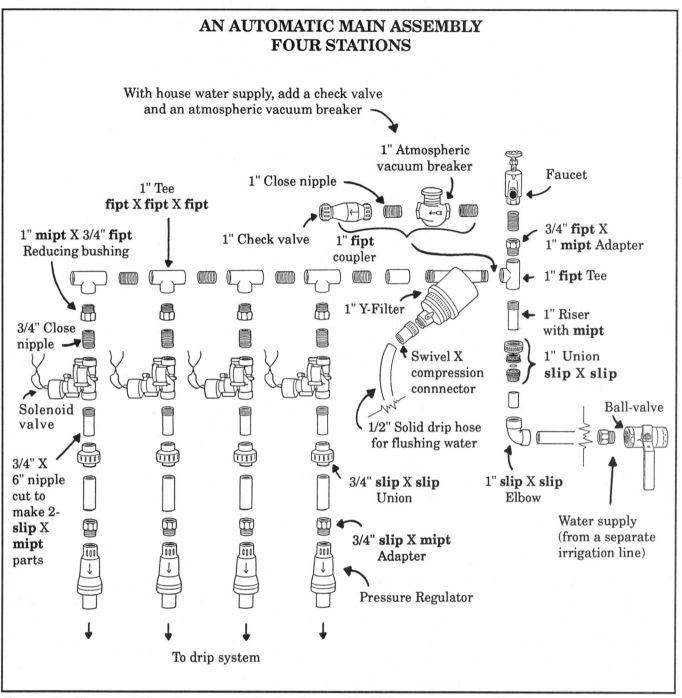

AN AUTOMATIC MAIN ASSEMBLY
FOUR STATIONS

With house water supply, add a check valve and an atmospheric vacuum breaker

1" Atmospheric vacuum breaker

1" Close nipple

Faucet

1" Tee
fipt X **fipt** X **fipt**

1" Check valve

3/4" **fipt** X 1" **mipt** Adapter

1" **mipt** X 3/4" **fipt**
Reducing bushing

1" **fipt** coupler

1" **fipt** Tee

3/4" Close nipple

1" Y-Filter

1" Riser with **mipt**

1" Union **slip** X **slip**

Solenoid valve

Swivel X compression connnector

Ball-valve

3/4" X 6" nipple cut to make 2- **slip** X **mipt** parts

1/2" Solid drip hose for flushing water

3/4" **slip** X **slip** Union

1" **slip** X **slip** Elbow

Water supply (from a separate irrigation line)

3/4" **slip** X **mipt** Adapter

Pressure Regulator

To drip system

Figure 45 All parts which make up a centralized main assembly for four separate zones. Each irrigation line to each zone has a solenoid 24VDC valve. The solenoid valves are automatically activated by an electronic controller.

Figure 45 shows what a centralized main assembly might look like. This one, for the west and north sides of the Shrubberys' house, supplies four lines, each leading to a separate drip irrigation system **(See Figure 46)**. All parts plumbed in after the water supply line have 1-inch threads because they will be used to irrigate several drip systems at once. The assembly begins with a ball-valve so the water supply can be shut off to work on any part of the assembly. Next comes a 1-inch union for easy removal of the entire assembly. The top "arm" of the 1-inch schedule 80 tee, just before the faucet, is stepped down to three-quarters of an inch with what's known in plumbing parlance as a reducer bushing (a 1-inch mipt X 3/4-inch fipt version) so that you can use a less expensive 3/4-inch brass faucet.

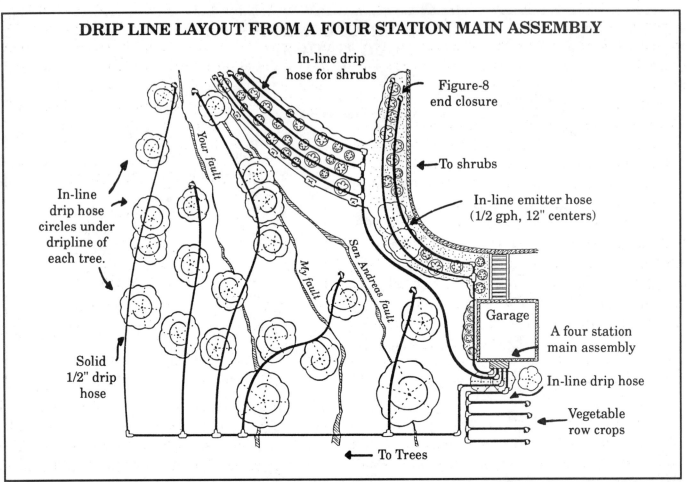

DRIP LINE LAYOUT FROM A FOUR STATION MAIN ASSEMBLY

In-line drip hose for shrubs

Figure-8 end closure

To shrubs

In-line emitter hose (1/2 gph, 12" centers)

Your fault

In-line drip hose circles under dripline of each tree.

My fault

San Andreas fault

Garage

A four station main assembly

In-line drip hose

Solid 1/2" drip hose

Vegetable row crops

To Trees

Figure 46 A view of the trees and shrubs at the Shrubberys' house; they are irrigated by the four valves of the main assembly. The large, well-spaced trees are serviced by a lateral of solid drip irrigation hose with a length of in-line drip irrigation hose circling beneath the dripline of each tree. Remember to lay out the tubing in a loose, slightly serpentine fashion so that as the tubing contracts in the fall it won't pull out of the fittings.

Off the "leg" of the first tee is where you add the atmospheric vacuum breaker and check-valve combo if you're connected to the house's water pipes. In this example, I've left out the anti-siphon devices because the assembly is connected to a water line from the well which services only the landscape, not the house.

A 1-inch Y-filter follows to allow for the simultaneous use of two or more irrigation valves. Everything up to this point is in the more expensive 1-inch fittings to allow for the possibility of running several drip irrigation lines at the same time. The "leg" of each 1-inch tee for each separate irrigation line is reduced to 3/4-inch threads with what's known as a threaded adapter bushing. This allows for less expensive valves and pressure regulators. And because each individual drip irrigation line usually doesn't require the flow of a 1-inch valve, you can revert to less

expensive 3/4-inch plumbing parts.

You can substitute any type of manual ball-valve for the automatic solenoid if you don't want the complexity or cost of an automatic system. (The threads are the same for both—fipt.) To join the solenoid (or any valve) to the union, cut a 3/4-inch, 6-inch schedule 80 nipple in half. Now you have two pieces, each slip X mipt. Use one to thread into the valve, with pipe dope, and glue into the slip X slip union. Be sure to use the special PVC glue for cementing schedule 80 plastic parts. The 3/4-inch unions with female slip fittings are added after each valve to simplify the removal of the entire assembly for any repairs.

If the flower bed or trees you'll be irrigating are nearby (less than 50 feet away) you can install the pressure regulators after the valves, at the main assembly. As always, paint the entire assembly.

But if there is any distance at all to the beginning of your irrigation lines, it is better to run schedule 40 PVC pipe from the main assembly to the beginning of the line and then add the pressure regulator (don't forget PMLP #6). If the pressure regulator were at the beginning of the system, the pressure drop from the friction of the water moving through the pipe out into the garden could severely reduce the flow. No matter where the pressure regulators are, be sure they have the right flow and psi ratings for each particular line or system.

When the regulator is at the end of the solid PVC supply pipe, it is connected with a male adapter, a mipt X slip 3/4-inch PVC fitting. Using pipe dope, thread the male adapter into each regulator. Then glue the male adapter to each solid schedule 40 PVC supply line. Finally, a 3/4-inch union with female slip fittings is glued (be sure to use the special glue for bonding white and gray PVC parts) into each line for simple removal of the assembly. As always, paint the entire main assembly.

Disregard PMLP #11 and bury the schedule 40 PVC lines at least 18 inches deep to protect them from future digging and cultivation. Leave the trenches open until all the below-ground parts have been glued together. Never bury the pipe until the glue has set for 12 to 24 hours and you've turned on the water to test for leaks. (A layer of sand placed around the pipe before backfilling with soil may help remind whomever is digging that her or she is about to hit a water line.) Use a 3/4-inch slip X slip elbow at the end of each PVC line to glue a piece of schedule 40 PVC pipe so that it reaches up a few inches above the soil's surface. Add another 3/4-inch slip X slip elbow above ground, facing toward your plantings. This is where you'll probably be adding the pressure regulator. Add a 3- to 4-inch length of PVC pipe parallel to the ground. Thread the male adapter (3/4-inch slip X mipt) into each regulator and glue the male adapters to each solid PVC supply line. Then the outlet side of the regulator is threaded with a mipt X mht transition nipple.

If your supply line doesn't have a regulator out in the garden, you can use slip X slip elbows to come up out of the ground and then connect directly to the drip hose. To easily adapt to drip irrigation fittings, glue a special male adapter, a 3/4-inch slip X 3/4-inch mht transition fitting (See Figure 47), to the horizontal PVC pipe coming out of the last elbow. Now simply thread the drip irrigation fht swivel X compression adapter onto the hose-threads of the transition fitting, insert the in-line drip hose and you're all set—if Pop Murphy hasn't meddled with your plumbing—to irrigate the modern, efficient way.

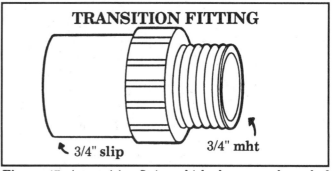

TRANSITION FITTING

3/4" slip

3/4" mht

Figure 47 A transition fitting which glues onto the end of PVC pipe and converts it to a male hose-thread.

The slip X mht fitting is not an easy part to find in an ordinary hardware store. Another method is to glue onto the vertical riser pipe an elbow which has female threads on one side—a 3/4-inch 90° slip X fipt (See Figure 48). Into the fipt threads add a mipt X mht transition nipple.

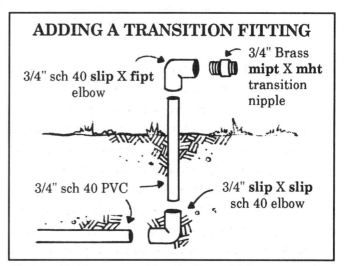

ADDING A TRANSITION FITTING

3/4" sch 40 **slip X fipt** elbow

3/4" Brass **mipt X mht** transition nipple

3/4" sch 40 PVC

3/4" **slip X slip** sch 40 elbow

Figure 48 Brass transition fittings which convert pipe-threads to hose-threads are easier to find in many hardware stores than a schedule 40 PVC slip X mipt fitting. Be sure to use Teflon pipe dope or tape to seal the threads. Any drip irrigation fitting with a fht swivel can be added to the transition fitting.

6 Drip Irrigation for Containers

Container plants—whether grown in expensive Italian terra-cotta pottery, plastic tubs, old olive oil cans, wooden planters or hanging baskets—add delightful accents to patios, porches, decks, windowsills and rooftops. For many urban gardeners, container plants are the only option for a soothing touch of foliage, color and fragrance, or even a modest harvest of fresh herbs, vegetables and some tree-ripened fruit. In all gardens, bedding plants in glorious full bloom, displayed in attractive pots, are a wonderfully easy way to provide colorful seasonal highlights throughout the garden.

Compared with their down-to-earth cousins, however, container plants are unforgiving of neglect, especially when it comes to watering. Pots dry out fast. Periodic and/or infrequent drenchings harm the plant by allowing intermittent periods of drought stress and by leaching valuable and limited nutrients from the soil. Daily watering with tiny amounts of water is often essential to maintaining consistent moisture and therefore consistent nutrient availability in the soil.

Enter drip irrigation, which fits container gardening like jelly on a peanut-butter sandwich. With a well-planned drip system you can add a small, measured amount of water to each plant every day. Consistent watering of container plants will reward the gardener with better growth and a longer period of bloom. Furthermore, drip irrigation can eliminate the mildew and fungus problems encouraged by overhead watering.

Drip irrigation of pots calls for a variety of hardware that I seldom use in landscapes. This is virtually the only situation where I mess with the dreaded "spaghetti" tubing. Its small diameter, either 1/8 or 1/4- inch, provides a low visual profile and makes it easy to hide among foliage. When carefully installed in small- to medium-sized containers, spaghetti tubing is manageable and doesn't end up the tangled mess so accurately suggested by its name.

Trickle Your Pot's Fancy

Start with the same main assembly that I've described for flower borders. Remember, if you use a combination check-valve and atmospheric vacuum breaker or just an atmospheric vacuum breaker, all the containers must be at least 12 inches *below* the level of the anti-siphon device. Only a more expensive brass-bodied reduced pressure backflow preventer with a double check-valve can service container plants or hanging baskets above the level of the main assembly.

Use the same quality of Y-filter, but make sure the stainless-steel screen is at least 150 mesh or up to 200 mesh. This smaller-sized screen will help prevent clogging of the smaller orifices of some of the emitters and emitter-like tubing used with pots.

Some of the parts you'll be using—Laser™ soaker tubing and porous pipe (sometimes called soaker hose)—should not operate above 15 psi and work best at 10 psi. Remember, quality pressure regulators are rated for a specific range of flow. You may not have many pots on any one container drip system and thus will be using only a very low volume of water. The Senninger low-flow pressure regulator, which comes in 10- and 15-psi models, is rated at a .1 to 8 gpm (6 to 480 gph) capacity. This is usually the best pressure regulator for containers, providing all the emitters or soaker hose in the system pass at least 6 gallons per hour.

Let's take a look at all the components for various sizes of pots, starting with a guide to matching the emitters to the container's size. Then I'll show you how, with a little sleight of hand, to make all the emitters and tubing virtually disappear.

Small Pots

For small pots, up to 12 inches in diameter, it's best to use a single emitter, no matter how many plants are squeezed in. To avoid overwatering small pots I find it best to use 1/2-gph emitters. If more water is needed I can always set the timer to stay on a little longer. As with landscaping, I always use pressure-compensating emitters for all my containers. While they might cost as much as 30% to 100% more than the noncompensating ones, I figure it's a small price to pay for the assurance that all emitters are dripping the same amount of water. The emitters will be attached to a main supply line (1/2-inch solid drip hose), using a length of spaghetti tubing. The barbed post on most emitters attaches to 1/4-inch spaghetti tubing, but a few emitters require the smaller 1/8-inch tubing. Be sure to buy the right spaghetti tubing for your emitter's barbed post.

Medium Pots

For medium-sized pots (12 to 18 inches in diameter), the rule is to use one emitter for every one to two plants. In a multiplant container you might get away with a single emitter with a higher flow rate, but if the water spreads unevenly through the potting soil, some roots may go unwatered and the soil may dry out and pull away from the sides of the container. Water could then drain out the sides without moistening the plant's roots.

If you don't want to load up the pot with emitters, there's another solution: a spot spitter. It's not as rude and crude as it sounds. The brand I use is the Roberts Spot Spitter™ because it is the most readily available and comes in the widest selection of flow rates and spray patterns. These spitters consist of a four-fluted plastic stake with a V-slotted post on the top and a solid, round post at the bottom (**See Figure 49**). Spaghetti tubing is inserted over the V-slotted post and water shoots out over the soil in a horizontal fan spray, which extends 12 to 18 inches, depending upon water pressure. The solid post on the bottom can be used to close the spaghetti tubing if the pot is empty. Simply lift the stake out of the soil, pull it off the spaghetti tubing, reverse the stake and insert the solid barb into the spaghetti tubing to plug it shut.

Spitters are not even closely related to emitters, as they distribute far more water. The width and depth of the V-slot in the top post determine, respectively, the pattern of the spray and its volume. At 10 psi, the amount of water distributed ranges from 5.5 gph for the 90° spray pattern to 9.3 gph for the 160° spray to 19.20 gph for the high-volume 360° spitter. (See the Spot Spitter Flow Rate chart in the **Appendix**.) Therefore, spitters cover a much larger area of the pot's soil surface than true emitters and add humidity to the atmosphere around a plant's foliage. One spitter can provide enough water for all the plants in a pot up to 18 inches in diameter. Make sure there is no foliage in the way of the spray pattern or you'll have dry spots in the potting soil. Spitters are best suited to containers which

are deeper than they are wide; this allows the soil to absorb the higher volume of water more readily and prevent it from running out the bottom.

It's best to use spitters with plants that prefer, or at least tolerate, high humidity and wet feet. Azaleas (*Rhododendron* sp.), rhododendrons (*Rhododendron* sp.), daphnes (*Daphne* sp.), ferns and sweet woodruff (*Galium odoratum*) are much more suited to spitters than are drought-resistant plants like rockrose (*Cistus* sp.), rosemary (*Rosmarinus* sp.), santolina (*Santolina* sp.) and California fuschia (*Zauschneria californica*).

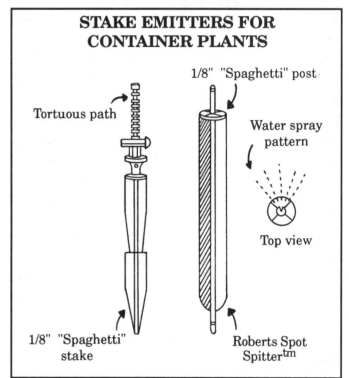

STAKE EMITTERS FOR CONTAINER PLANTS

Figure 49 Two types of "spaghetti" tube stakes, the tube over the top post. The one on the left has a tortuous path "emitter" built into the top of the stake. The right-hand one is a spitter, which makes a small spray pattern.

Installing Drip In Small and Medium Pots

The most invisible drip system for pots is installed as the containers are planted. To begin, lay a length of 1/2-inch solid drip-hose from the main assembly near to the base of all the pots. You must bring the 1/2-inch tubing close to the pots and not rely upon spaghetti tubing to travel a long distance. As listed in the Maximum Length of Tubing chart in the **Appendix**, the longest possible length for consistent and adequate flow with 1/8-inch spaghetti is 5 feet, and

15 feet for 1/4-inch spaghetti. The solid drip hose can be, in spite of PMLP #11, in a trench alongside a patio, under the wooden flooring of a deck, or hidden beneath the handrail of a porch. After the drip hose is in place, turn on the main assembly and allow the hose to flush any dirt out the open end. Turn off the water and close the end of the drip hose with either a figure-8 end-closure, a compression end-closure with a thread cap or a Spin Loc end cap.

Before adding the spaghetti tubing and planting your pot, be sure to seal the inside. Terra-cotta pots are especially prone to wicking lots of moisture through their walls. I brush an inexpensive, nontoxic cement sealer on all the inside surfaces of clay pots. Sometimes the clay soaks up so much sealer that two coats are required.

At each pot, punch a hole in the solid drip hose and insert a transfer barb—which has two barbed posts, one to go into the drip hose and the other to insert into the spaghetti. Be sure to use a transfer barb which is the same size as the spaghetti tubing. Use spaghetti tubing to go from the drip hose to beneath the pot (**See Figure 50**). Cut a piece of spaghetti 4 to 6 inches longer than the soil depth will be in the pot, attach one end to the spaghetti elbow beneath the pot and thread the tubing through the hole in the bottom of the pot before you add any potting soil. If your pots are on a patio or rooftop, you'll need to block each one up off the surface a bit to allow the spaghetti to come up through the bottom of the pot without getting flattened. With wooden decks, simply set the drainage hole of each pot over the space between two boards.

Next, use a single piece of broken pottery or a flat pebble to close off any space around the spaghetti where it comes through the drainage hole; this will keep any potting soil from falling out. Now you can fill the container with a rich, well-drained potting soil and plant. Leave the spaghetti tubing above the soil open and flush it for several minutes to clear any soil out.

Now trim the tubing to fit and add the appropriate emitter or spitter to each length of spaghetti. You can use a spaghetti stake to hold the emitter in place. Be sure to keep the emitter away from the base of the stem or trunk to prevent crown rot. If you need more than one emitter in the pot, use a spaghetti barbed tee to branch out for two emitters. Or have more than one length of spaghetti coming up through the bottom of the pot, each attached to its own transfer barb. A little bit of mulch will further disguise the tubing and emitter. Even the day you plant, your container drip system should be practically invisible.

If any one pot in a system should become empty, you can save water by pulling the emitter or spitter off the spaghetti tubing and inserting the right-sized barbed goof plug.

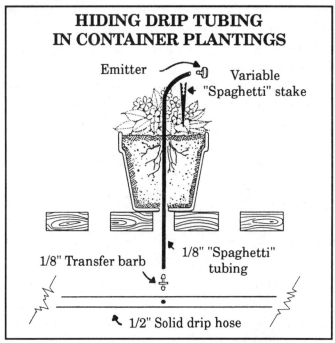

HIDING DRIP TUBING IN CONTAINER PLANTINGS

Emitter — Variable "Spaghetti" stake

1/8" Transfer barb — 1/8" "Spaghetti" tubing

1/2" Solid drip hose

Figure 50 Drip irrigation of containers can be practically invisible. Attach a length of "spaghetti" tubing to a lateral of solid 1/2-inch drip irrigation hose, thread it up through the pot, plant and secure the emitter at the end of the tubing with a "spaghetti" stake.

Large Pots

You could use a number of punched-in emitters in pots over 18 inches in diameter, but there are two better solutions—porous pipe and Laser tubing. Both can be coiled in rather small circles to provide a continuous ring of water to moisten all the potting soil evenly.

The easiest of these to use is 1/4-inch Laser tubing, which has tiny laser-cut slits every 6 inches in tubing that closely resembles regular 1/4-inch spaghetti tubing. This tubing passes 1.2 gph per lineal foot. Notice that the Laser tubing has small white arrows printed on it. Don't forget PMLP #6; make sure the arrows point away from the source of the water and toward the end of the tubing. Plant and install your pots as described above, leaving only a few inches of solid 1/4-inch spaghetti tubing above the potting soil (**See Figure 51**). Flush the system. Next, clip the spaghetti off just above the soil and attach a 1/4-inch barbed spaghetti elbow (90°). Add a length of 1/4-inch Laser tubing to the elbow and circle the pot just a few inches in from the perimeter. With large pots you can coil the Laser tubing in a spiral pattern, but make sure you don't water directly on top of the crown of the root system. You can hold the tubing in place with metal J-stakes every foot or so. Flush the tubing clean and close the end by inserting a 1/4-inch barbed goof plug.

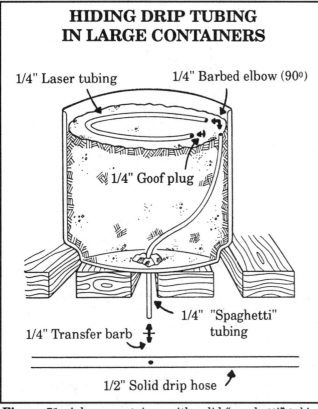

HIDING DRIP TUBING IN LARGE CONTAINERS

1/4" Laser tubing   1/4" Barbed elbow (90°)

1/4" Goof plug

1/4" Transfer barb

1/4" "Spaghetti" tubing

1/2" Solid drip hose

Figure 51 A large container with solid "spaghetti" tubing coming up through the soil. A ring of Laser tubing lies on the soil's surface to evenly water the entire container.

As noted in the Maximum Length of Tubing chart in the **Appendix**, 1/4-inch Laser tubing should never be used in lengths of more than 15 feet per lateral or, in this case, per pot. If your pot or container is so large that more than 15 feet of tubing would be required, you can switch to 3/8-inch porous pipe or 3/8-inch Laser tubing.

As mentioned earlier, porous pipe is constructed with a labyrinth of tiny passages throughout its walls. As a result, the entire surface oozes or sweats water at a rate of .6 to .8 gph per foot. The pipe is pliable enough to coil around pots 18 inches or bigger. This material shouldn't be used with water that is high in iron, calcium or other dissolved minerals or that is unchlorinated, (which may allow for an algae bloom in the tubing), as in all these cases the interior labyrinth can become clogged, even when a fine-mesh filter is used. Be sure you have a 10-psi regulator in the main assembly to keep from blowing the porous-pipe insert fittings apart, a manifestation of PMLP #10 *and* #4.

To the solid 1/2-inch solid drip-hose supply line add a 1/4-inch X 3/8-inch transfer barb; these are sold as Laser tube adapters. (See the middle pot in **Figure 52**.) Punch the 1/4-inch barb into the solid 1/2-inch hose. Add a length of solid 3/8-inch drip hose to come up through the planted pot. (They don't make a 3/8-inch barbed elbow, so you may want to use 1/4-inch spaghetti tubing and barbed 1/4-inch elbows to an elbow above the soil and then add the 1/4-inch X 3/8-inch transfer barb.) Once the solid tubing is out of the soil, add a 3/8-inch barbed coupler and the required length of 1/2-inch porous pipe, which has an internal diameter of 3/8-inch. You may have to soak the porous pipe in hot water to fit it over the 3/8-inch barb. Hold the pipe into the desired position by pinning it every 12 inches or more with U-shaped metal pins, called landscape staples or U-stakes, which come in 6- and 9-inch lengths. I always use the longer ones because they hold the tubing better in loose potting soil. Flush the porous pipe. Close the end of the coiled porous pipe with a Laser air release end cap, which inserts into the open end of the porous pipe. You'll more easily circumvent PMLP #4 and have a sturdier system, less prone to leaks, if you add a #8 (1/2-inch or less) stainless steel hose clamp at every point where a fitting inserts into the porous pipe. Finally, mulch the container to cover the larger, more noticeable, porous pipe.

If your plant is already potted or the drainage hole is too small, you'll have to bring the water in over the pot's lip. While this is far more noticeable, it is the only choice. You'll have a better water supply if you bring 1/2-inch solid drip hose over the back lip of the container using compression elbows (See the right-hand pot in **Figure 52**). Add a short length of 1/2-inch hose to extend over the lip, flush and cap. Then insert a 1/4-inch X 3/8-inch transfer barb into the hose and add 5/8-inch porous pipe as described above. Don't forget PMLP #4—use hose clamps.

Hanging Planters

Hanging plants present a special consideration because they are even more prone to dehydration than earthbound containers. You can use regular emitters in hanging pots, but the pots usually need more water than the same-sized pot on the ground. So if you're using, for example, a single 1/2-gph emitter in a 10-inch pot on your patio, try at least two emitters or one 1-gph emitter in the same-sized hanging pot.

Spot spitters put out a good amount of water over a large area, but the flow rate is so great that the water may run off the surface or down the sides of the pot faster than it can percolate into the potting mix. I don't recommend them in most hanging plantings.

Many hanging plants are water and humidity lovers. While a spitter does produce some humidity, it's nothing compared with the fog-like vapors of a micro-mister. While they pass more water than some of the lowest-flow emitters, misters use only 3 gph to generate a delightfully cool cloud. With the mister placed beneath the plant's foliage, the moist fog condenses on the plant's leaves and drip-irrigates much of the pot's surface, although some water will dribble off the foliage outside the pot. Two name brands for 3-gph misters are Hit ProMist™ and the Hardie Mini-Mister™ (**See Figure 17**). Each comes with a special stake to hold the mister upright so the spray billows up into the leaves. The Hardie Mini-Mister costs less and its stake and mister are considerably less bulky and therefore not as noticeable as those of the Hit ProMist. The use of these

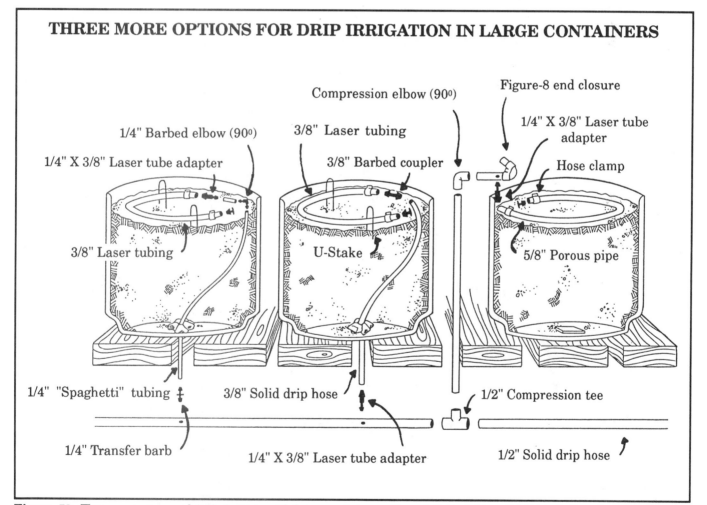

THREE MORE OPTIONS FOR DRIP IRRIGATION IN LARGE CONTAINERS

Compression elbow (90⁰)

Figure-8 end closure

1/4" X 3/8" Laser tube adapter

1/4" Barbed elbow (90⁰)

3/8" Laser tubing

Hose clamp

1/4" X 3/8" Laser tube adapter

3/8" Barbed coupler

3/8" Laser tubing

U-Stake

5/8" Porous pipe

1/4" "Spaghetti" tubing

3/8" Solid drip hose

1/2" Compression tee

1/4" Transfer barb

1/4" X 3/8" Laser tube adapter

1/2" Solid drip hose

Figure 52 Three ways to use drip irrigation with large containers. From left to right, solid 1/4-inch "spaghetti" services a ring of 3/8-inch Laser tubing, 3/8-inch solid hose is attached to a circle of 3/8-inch Laser tubing and 1/2-inch solid drip hose comes up behind the container and over the lip to provide water to a coil of 5/8-inch porous pipe.

misters requires a 200-mesh filter screen to prevent the tiny misting hole from clogging.

Hanging plants are often high enough off the ground that 1/4-inch spaghetti, with its 15-foot limit, won't reach. In this case, 3/8-inch solid drip hose can be laid out for up to 100 feet and is much less visually bulky than 1/2-inch solid hose. Use 3/8-inch tubing (with an outside diameter of .455 inch) and the appropriate number of compression elbows and tees (called series 400 fittings) to get as close to each hanging plant as possible. The tubing is cut and wiggled into the compression fittings in the same way you put together in-line emitter hose. Try to hide the 3/8-inch tubing beneath the batten of a board-and-batten wall, in a routed notch up the backside of a post, or beneath a railing or eaves. Use metal pipe-straps to secure the tubing flat to the post or wall. It's important neither to fasten the tubing securely to the post nor to stretch the hose taut, because winter's cold will significantly shrink the tubing and may damage fittings **(See Figure 53)**. Paint the tubing to protect it from ultraviolet light and to help it blend in with the surroundings. At each hanging pot, use a 1/4-inch transfer barb to attach a length of spaghetti tubing. Run the spaghetti down one of the wires which holds the pot, tying the tubing at intervals to keep a low visual profile. Flush the system and add the appropriate emitter, spitter or mister.

Large rooftop planters or beds have a proportionately larger soil volume and are less vulnerable to rapid drying. They can easily be irrigated with lengths of either 1/2-inch porous pipe or in-line

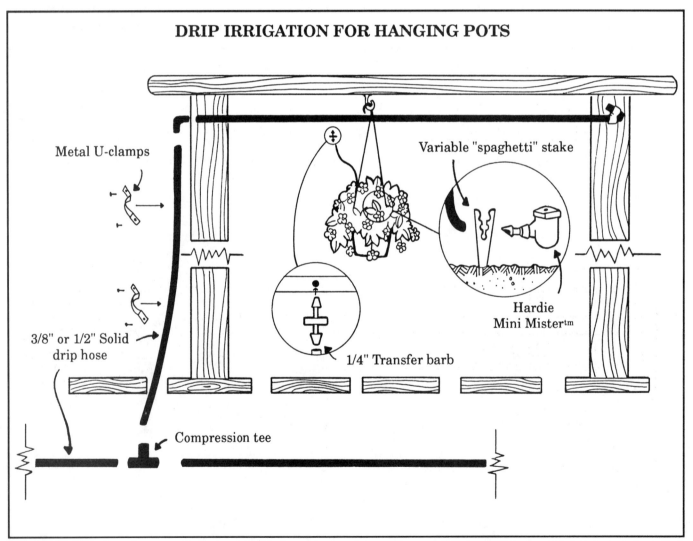

DRIP IRRIGATION FOR HANGING POTS

Metal U-clamps

Variable "spaghetti" stake

3/8" or 1/2" Solid drip hose

Hardie Mini Mister™

1/4" Transfer barb

Compression tee

Figure 53 Hanging plants can be irrigated with low-volume misters. Three-eighths- or 1/2-inch solid drip hose is installed up the arbor's post and above the hanging plants. One-quarter-inch spaghetti tubing services a micro-mister mounted on a spaghetti stake.

emitter tubing. I irrigate such beds in the same manner as I would a wooden-sided raised bed in the garden. This is discussed in Chapter 8, Drip Irrigation for Vegetable Beds.

Segregating Your Drip System

Notice that in the drawing of Mr. and Mrs. Shrubbery's yard, there are several separate drip systems for different potted plants **(See Figure 54)**. Their yard, like many landscapes with a mixture of container plants, can be serviced with drip irrigation, but you can't always hook up everything to the same system.

The reason for this lies in the flow rate. A drip system with 1/2- to 1-gph emitters cannot easily include high-flow devices such as spitters, Laser tubing and porous pipe, which have gph ratings ranging from several gph to nearly 20. Irrigating long enough for the emitter-watered plants would waste a lot of water on the spitter or porous pipe irrigated ones and could even drown them.

For example, if you want to combine the installation of emitters in small pots with the use of

lengthy pieces of Laser tubing coiled up within big pots, you'll have to run them off two separate faucets. Or you could plumb a Y-valve (with two small ball-valves) after one filter and add two different pressure regulators, one at 25 psi for the emitters and the other at 10 psi for the Laser tubing. Each side should be turned on and off independently to allow for different irrigation rates. While you will usually water each side of the Y-valve for a different length of time, you may occasionally have both lines on at the same time. Therefore, make sure the total flow of the two lines doesn't exceed the capacity of the filter.

Hanging plants should also always be on their own valve with a timer because their water needs are so special. If you live in an especially hot or windy area, get a timer which allows you to turn on the system automatically several times during the same day to keep hanging plants well watered and healthy.

Container plantings usually demand daily irrigation so that the soil doesn't dry out and pull away from the walls of the vessel. If this happens, subsequent irrigations can simply run off the soil surface and down the cavity between the soil and the planter to drain

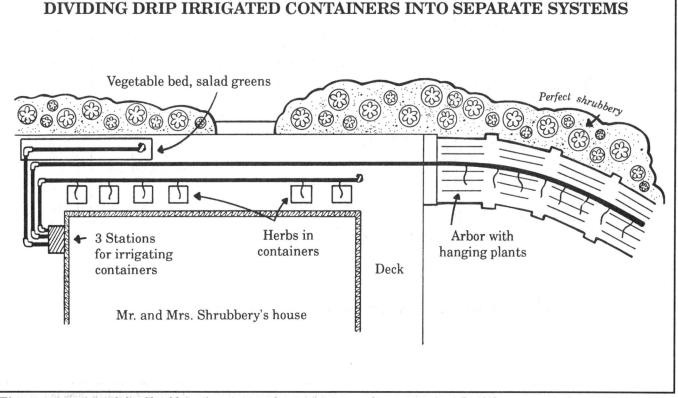

DIVIDING DRIP IRRIGATED CONTAINERS INTO SEPARATE SYSTEMS

Vegetable bed, salad greens

Perfect shrubbery

3 Stations for irrigating containers

Herbs in containers

Deck

Arbor with hanging plants

Mr. and Mrs. Shrubbery's house

Figure 54 A view of the Shrubberys' container plants. A separate line irrigates each of three areas: a long raised bed for vegetables, six large containers on the back porch for culinary herbs and a number of hanging plants in the arbor.

away unused. Slight amounts of daily water keep the soil mixture moist, plump and in full contact with the walls of the container. The daily amount of water added should equal what is lost by evaporation from the soil and container and by transpiration from the plant's leaves. By adding just enough water to equal these losses, you will also prevent leaching of the limited and valuable nutrients in the potting mixture.

Only experience and experimentation can provide the specific answer to "How long do I leave the drip system on?" Start with very short times of only a minute or two. If you're going to automate your container irrigation, beware—not all irrigation timers or controllers can be set for times as short as one or two minutes. Check with your supplier to make sure you purchase a timer with short "run" times. Observe how the soil and plant in each container respond during the first several days and then adjust the time up or down, depending upon your best evaluation.

In my own garden, I've used drip irrigation in some rather narrow but deep containers (some with their bottoms open to the ground) and some medium-sized terra-cotta pottery along the south wall of my house, all emitters controlled by one battery-operated timer. The emitters in each container are rated at only 1/2 gph. During the summer, when the eaves of the house block much of the midday sun, I set the timer for only 3 minutes—a tiny amount of water from a 1/2-gph emitter. When the heat of Indian summer hits the containers full blast, I increase the time to 5 or 6 minutes—still only about one-twentieth of a gallon. In late fall, the sun drops behind a number of evergreen trees and provides only limited filtered light for the duration of the winter and early spring. For all of the winter, except when dry spells have been particularly prolonged during California's recent six-year drought, the timer is turned off and the periodic rains take care of irrigating the containers.

7 Drip Irrigation for Trees and Shrubs

On a sloped, east-facing hillside in Sebastopol, CA, grows a remarkable orchard that produces a cornucopia of fruit while serving as a testing ground for many types of drip irrigation hardware. Sponsored by the University of California Cooperative Extension; Roy Germone, who provided the land; and the Master Gardener Program (which is sponsored by the Cooperative Extension) and several local businesses which have provided raw materials and services, this unique demonstration plot features 200 fruit trees and dozens of cane and bush fruits. Each row utilizes a different type of drip irrigation emitter or micro-sprayer (tiny sprinklers similar to large rotating lawn sprinklers, such as the Rainbird™ impact sprinklers). This has allowed Paul Vossen, local UC Cooperative Extension tree crops farm advisor, to test many possible kinds of drip systems for fruit trees simultaneously, evaluate their performances for both home and commercial plantings and demonstrate to both home gardeners and farmers how this hardware might work for them.

The trees at the Germone Demonstration Orchard have been showing some impressive results. Paul's records during the last six years show that "with some of our peach trees, we're getting 14 tons per acre in the third year while local growers are only getting seven tons, and on three-year-old apple trees we're getting 20 tons per acre, while established, unirrigated, and mature apple orchards are only getting 13 tons per acre." (These impressive increases in yields correspond to an increase in abundant foliage and plentiful bloom produced on ornamental trees when appropriate irrigation is used.)

Drip Irrigation for Good and Healthy Root Growth

To get such superior yields, one needs to know how roots grow. Paul begins with his knowledge that contrary to popular belief, "the top two feet of soil represents 80 percent of the tree's root mass. Therefore most of the water and nutrients absorbed by the trees come from this zone in part because that's where the oxygen is to fuel the soil biota that provide the nutrients." Paul maintains that "the roots will develop

where there is water, but they can't grow any distance at all from a dry spot toward moisture. The tree's roots don't *explore* for water, they *exploit* available water."

The Germone Demonstration Orchard trees were planted on top of 18-inch-tall, 4-foot-wide berms that run the entire length of each row of trees. So as not to counteract the effect of a dry, well-drained berm, Paul reasoned that the irrigation system should keep the tree's root crown and trunk dry at all times. Conventional sprinklers, especially large Rainbird-type high-impact sprinklers, wet the entire surface of the orchard, the root crown, the trunk and even the tree's foliage—thus encouraging diseases of the roots, trunk and leaves.

This study has shown that drip irrigation not only avoids all the potential drawbacks of conventional sprinklers but also keeps the top 2 feet of soil at optimal moisture levels and reduces weeds. Furthermore, drip irrigation puts less demand on the orchard's well pump and makes better use of the water supply.

Each row of trees has one type of drip irrigation emitter or micro-sprinkler installed along its entire length. In all, there are working demonstrations of seven types of emitters, one version of drip tape, and five kinds of micro-sprinklers. Each row of fruits has a single line of 1/2-inch solid drip hose running down the length of the row near the base of each tree, on top of a raised planting berm. The emitters are punched into the drip hose at 18- to 24-inch intervals on either side of each tree's trunk to keep the drip irrigation applied water away from the root's vulnerable crown. The raised soil berm allows for plenty of drainage to protect the upper portion of the root system from crown rot (*Phytophthora* sp.). There is no hosing in the flat area between berms; this allows for periodic mowing of the permanent ground cover of either Dutch white clover or subterranean clover and Zorro fescue.

For extra protection, every set of three drip irrigation lines (three rows of trees) has a 200-mesh metal screen filter in addition to the filter at the well. Most home owners, with smaller plantings than the Germone Demonstration Orchard and with cleaner water or city water, will need only a 150-mesh filter.

Every set of three rows is controlled by a 24VAC electric solenoid valve. Each solenoid valve is in turn automatically controlled by a circuit, usually called a "station," on an electronic microchip irrigation controller. This orchard requires a seven-station electronic irrigation controller, which controls the irrigation start and stop times to all 20 rows—this includes tree crops and bush and cane fruits. (This summons up PMLP #12, but that is discussed in Chapter 10, Controlling Your Drip Irrigation.)

Evaluating Drip-Irrigation Emitters

Conventional thinking has held that emitters with lower flow rates such as 1/2-gph and 1-gph emitters, are more likely to clog and that the larger 2- to 4-gph emitters are less likely to become clogged with silt. Paul has all sizes in the Germone Dmonstration Orchard, and, with the exception of the 4-gph emitter, emitters for each given flow rate have come from several different manufacturers.

According to Paul, "The 1/2-gph emitters didn't plug up on us very much. They did plug up more than the 2- and 4-gph emitters, but not very much at all." Some of the 1/2-gph emitters had approximately 20% clogging, due to the high level of iron in the well water. The 1- and 2-gph emitters; with the name brands Spot Vortex™, DBK™, Netafim™, and Solco™, worked well.

Drip Irrigation for Landscape Trees

The Germone Demonstration Orchard is a model for conventional row-based orchards.

Now, returning to our ideal drip-irrigated garden, Mr. and Mrs. Shrubbery have chosen, as seen in **Figure 20**, to arrange their fruit trees in the more informal style of an edible landscape. The trees are grouped together because they have similar water and fertilization needs. The layout, however, is more like a forest, with random spacings and clusters of certain trees, which puts the pollinators next to each other. The entire area is mulched with a substantial layer of shredded trimmings from the local tree service. This thick layer of chipped bark and shredded leaves serves a number of purposes: it hides the in-line emitter drip hose, conserves moisture, keeps the sun from overheating the soil, prevents wind-blown soil erosion,

provides safe and easy footing for pruning and harvesting the trees and makes fallen fruit easily visible for a thorough cleanup to prevent the wintering over of pests and disease.

Irrigating All of the Tree's Feeder Roots

This design allows for a layout even more efficient than that of the Germone Demonstration Orchard. The limitation of conventional drip irrigation in a typical orchard is that it consists of only a single row of hose with emitters, running straight down a row of trunks. This layout limits the water to a relatively small proportion of the tree's root system. Generally speaking, the roots of a tree actually extend one-and-a-half to three times farther than the width of the foliage. In heavy clay soil the roots are limited to an area one-half again as wide as the foliage. Sandy soils offer much less resistance, and the roots will often grow three times wider than the branches aboveground (See **Figure 55**).

The roots responsible for absorbing water and nutrients are the very young and tiny root hairs found near the tips of new roots. Older roots, much like older branches, have a bark-like covering which protects the root but doesn't produce root hairs. In one study done with radioactive isotopes in England, the roots 4.5 feet away from the trunk of a 10-year-old apple tree absorbed just less than 10% of the water and nutrients absorbed by the entire root system. The study showed that proportionately, many more feeding roots are found at, or beyond, the dripline of a tree's foliage. Because of this, I prefer to place a ring of in-line emitter tubing at or beyond the tree's dripline. This produces a "doughnut" of moisture, and far more of the root zone is adequately irrigated than with a single line of hose down a row of trees (See **Figure 56**). As the tree grows, extra lengths are added to this circle of moisture to correspond with the new growth in the canopy. By setting the circle of emitters farther out every few years, you're encouraging the tree's roots to explore more soil volume and gather more nutrients for superior growth. Other than placing parallel rows of in-line emitter tubing, like that in a flower border, this is the most efficient way to irrigate a large volume of the tree's roots.

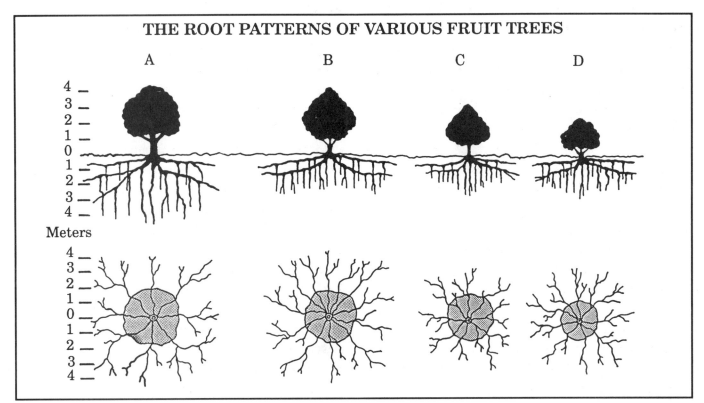

THE ROOT PATTERNS OF VARIOUS FRUIT TREES

Figure 55 Aboveground and root systems of 13-year-old fruit trees growing in the orchard of the Timiryazev Agricultural Academy, USSR: A–apple; B–pear; C–plum; D–sour cherry. The shaded circles drawn on the top-down view show branch spread; the black lines are the rooting pattern.

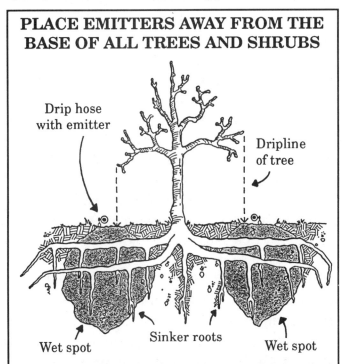

PLACE EMITTERS AWAY FROM THE BASE OF ALL TREES AND SHRUBS

Figure 56 Both young and established trees and shrubs generate most of their feeding roots at or beyond their dripline. To protect the root's crown and supply moisture to the largest number of root hairs, place the emitter at or beyond the dripline of each tree. The underground wet spot will moisten a large volume of soil.

To get from tree to tree, use solid 1/2-inch drip hose (with an outside diameter of 16mm, or .620 inch.) Placing the emitters as far away as the dripline of the newly planted or established shrub or tree is necessary to protect the upper portion of the root system from moisture-induced rot. The wide wet spot spread below the soil surface by the emitters guarantees that the young feeder roots will receive moisture. At the dripline of each tree, use a Spin Loc tee to add the correct length of in-line pressure-compensating emitter hose. After all the trees have been encircled, turn on the main assembly for 5 minutes or more to flush the hosing. Use a figure-8 end-closure (or an end cap) at the end of each in-line hose **(See Figures 57 and 58).** Start with the tree closest to the main assembly. Close each tree's in-line hose in sequence until you've worked your way to the end of the assembly. You can keep the hose in place with a U-stake every 4 to 6 feet. When it comes time to add another length of in-line hose, simply remove the figure-8, cut off the crimp in the hose just an inch or so toward the main assembly, add a compression or Spin Loc coupling, attach the new length of hose, flush and close the new end with the figure-8 end-closure.

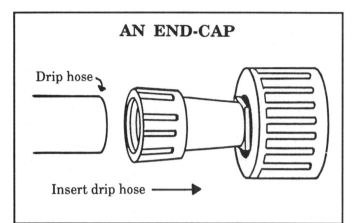

AN END-CAP

Drip hose

Insert drip hose ⟶

Figure 57 A Spin Loc end cap. Unscrew the fht cap, on the right side of the illustration, to flush the drip hose periodically.

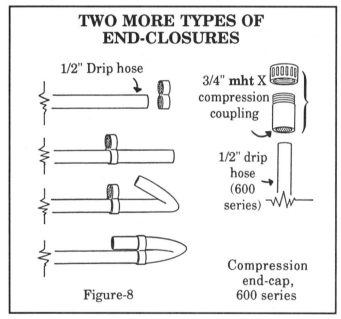

TWO MORE TYPES OF END-CLOSURES

1/2" Drip hose

3/4" **mht** X compression coupling

1/2" drip hose (600 series)

Figure-8

Compression end-cap, 600 series

Figure 58 Two more ways to close the ends of each line of drip hose. The drip hose will eventually crack with repeated use of the figure-8 end-closure.

Remember, the total length of in-line emitter hose per valve or main assembly has to be 326 feet or less. If your trees are very far apart, the solid drip hose which runs from tree to tree should total less than 240 feet.

This same process of encircling a tree with a doughnut of moistened soil works fine with ornamental trees. Shade trees in particular respond with excellent growth and a fast-growing, shady canopy. Ornamental flowering trees will have superior growth, enhanced blossoming and an extended length of bloom.

How Much Water?

It was from Paul Vossen's remarkable flyer titled "Water Management for Fruit Trees and Other Plants" (published in 1985) that I first learned about ET-based irrigation rates. The publication begins with a provocative question: "Did you know that a large apple tree, on a hot summer day, will use about 50 gallons of water?" That's right, 350 gallons per week for each individual, mature ornamental or fruiting tree with a canopy that covers 300 square feet of the ground.

This small blue leaflet of Paul's caused quite a skeptical stir among gardeners, landscapers and farmers back in '85. But after a thorough review and several years of applying Paul's suggestions, I realized that the "Daily Water Use (ET)" chart was one of my most valuable irrigation "tools."

Remember, the ET rate varies from place to place, season to season, and even day to day, but once you know the ET rate (expressed in number of gallons per day or per month), you know, more accurately than by any other method, how much water to apply. To find out the ET rate for your area, by month or season, call the local Cooperative Extension, the local library's reference desk or a local university, college or junior college horticulture department. Get the monthly ET rate, since it can vary considerably from the cool days of April to the dog days of August.

Paul's guide, "Daily Water Use (ET)," which is the basis for **Figure 30**, gives you the irrigation rate for replacing what is lost each day or month, but you can also increase the irrigation rate from this basic figure if you want to increase growth and yields. Because the ET rate is the only "universal" guideline, no longer should a gardener in Texas have plants die after reading an article by a New England garden writer who casually and misleadingly says to "water every 10 to 14 days till the ground is wet." Nor will a New England reader drown plants by using the rule-of-thumb watering rates proposed by a California garden writer.

Once you have the ET rates for various months, or an average one for the entire growing season, refer to **Figure 30**. Notice how the ET rate is listed in inches per month across the top—1 through 10 inches. Match the number you got for your local ET and read down the column. Within the body of the chart, the numbers represent gallons (or a fraction thereof) of water *per day*. The left-hand column represents the square footage of the canopy or foliage cover. It is important to focus on the square footage, since a plant's size at any given age can vary. Paul cautions the reader to always consider the amount of ground covered by *all* foliage. "Percent of cover," he cautions, "pertains to *everything* that's growing on the area that's to be irrigated. If you have young fruit trees far apart, but there's a perennial ground cover over the whole area—then, you'll have to water based upon solid cover rate to keep both the tree and the perennial ground cover happy." For such a planting, divide the 1 acre solid cover rate by the number of square feet in your planting or landscape to get the daily water use for a given ET rate. (Or multiply the area of your planting by the 1-square-foot figure at the top of the chart.) Since this is such an important concept and is so often ignored or misunderstood, Paul always adds another reminder—"Water is water, and roots is roots. If it's green, it's using water. Base your irrigation upon the ET rate and the area of *total* foliage cover."

The number you get from the chart is really just a starting point. The amount can be increased quite a bit to make growth and yields go up correspondingly. For example, according to **Figure 30**, with an ET rate of 6 inches per month, my newly planted Asian-pear orchard, with each tree's foliage covering 4 square feet, needed only .5 gallon per day and 3.5 gallons per week per tree. I decided to water a larger area around each young tree and let the natural grass ground cover grow. For the first year, I applied 35 gallons per week per tree, and I was amazed. The trees put 3.5 to 4.5 feet of new growth on four or more primary branches per tree yet the trees didn't appear overly watered or too weak or spindly. This increase in watering follows in the footsteps of Paul's recommendation to create "water junkies." "Don't hold back on the water when your trees are young," he advises, "you've got to build a well-branched and extensive branching to the trees *before* they come into fruiting." This also applies to ornamental trees for shade and blossom.

As with herbaceous flowers, you can water daily (for the best growth) or weekly, or even monthly, providing the amount is based on the ET rate. To determine the length of each irrigation, start by totaling the number of emitters built into the in-line hose in a system and multiply by the flow rate of the emitter to determine the system's total capacity in gallons per hour. To determine how many hours to leave the system on, divide the number of gallons recommended in the ET chart by the total gallonage the system can distribute.

For example, a tree covering 300 square feet of ground might have 11 lines (1.5 feet apart) of in-line pressure-compensating hosing, each 18 feet long, for 198 total feet of in-line hose. Since the emitters are every 12 inches, the total number of emitters is also 198. With a rating of 1/2 gph, the entire system has the capacity to distribute 99 gallons per hour. Where the ET rate is 3 inches per month, 300 square feet of canopy requires at least 18.6 gallons per day. Dividing 18.6 gallons, the plant's water needs, by 99 gallons per hour, the system's total capacity, we get .19 hour. Leave the system on for a little more than 11 minutes (.19 times 60 minutes) each day to equal the ET rate for 300 square feet of foliage. This can be adjusted up or down, based upon your observations and needs.

Drip Irrigation of Shrubs

On the south side of Mr. and Mrs. Shrubbery's house, they have an area of . . . well, how else can I say it. . . shrubbery. Most shrubs are planted close enough together that their roots, if given adequate irrigation, are all intertwined. Instead of the complications of circling in-line hose around each shrub, I simply irrigate the entire area as described for the flower border. Parallel lines of in-line hose are placed every 12 inches apart in well-drained soils and up to 24 or 30 inches apart in heavy soils—as per your initial test with the emitter punched in a plastic milk jug or your personal experience with the flower border. Be sure to jog the in-line hose around the base of each shrub to protect the root's crown from rot.

Again, base your irrigation schedule on the appropriate ET rate.

The trick to drip-irrigating vegetable beds is to make the system as simple as possible and easily adaptable to routine tasks such as weeding, hoeing, digging and fertilizing. For many years I never recommended drip irrigation for vegetables to my clients because I couldn't figure out how to make the system both safe from damage *and* convenient.

Early "experiments" in vegetable drip irrigation by my gardening friends started with the use of Mono-tubing™ or Bi-wall™. Mono-tubing™ is a single thin-walled plastic tube with minute holes every 12 inches, much like the old-fashioned soaker hoses my grandfather used, which were flat and green with white stripes. Mono-tubing is so thin-walled it collapses flat like a long balloon after each irrigation. While not exactly balloon-thin, Mono-tubing is extremely easy to pierce or tear; it practically splits open if it sees a hand trowel coming. Furthermore, the minute holes are prone to clogging, even when proper filtration is used. Bi-wall is made with a bit heavier-gauge plastic tubing and with a second, smaller chamber or tube, which has tiny holes spaced every 12 inches, piggy-backed on top of the larger supply tube. The second chamber helps to moderate the pressure along the length of the line, but it too is very prone to clogging, even with a 200-mesh screen in the filter. In both cases, due to their propensity to clog, the tiny holes need to be placed facing up. These two products don't really drip, like the old-fashioned soaker hoses, but rather make small, thin streams which arch 18 inches or more into the air, depending upon the pressure. This, of course, negates one of the greatest advantages of drip irrigation, that of keeping moisture off the foliage of vegetable plants such as squashes, cucumbers, melons and tomatoes, which are prone to water-induced mildew, rust and fungus. T-Tape™ is a more recent product, is similar to Bi-wall but has the advantage of not producing the arching streams of water.

I think the original fascination with these three products was the low cost of the materials. While this is a major consideration to farmers with acres of fields, it's of less importance to all but the poorest home gardeners. All three tubings must be carefully stretched in straight lines, even secured at the far end with a stake, to prevent twisting and kinking, which can cut off the water's flow. They cannot be adapted to the gentle curves so often found in landscaping, nor can they be easily plumbed to make 90° turns. While some of the material is rated for more than five years of *careful* use, most grades are damaged within three years. None can compare with the factory-rated guarantee of ten years for both solid and in-line drip hosing. Currently, T-Tape and Bi-wall are used by farmers. Most gardeners have switched to drip hose or porous pipe.

I have, in the past, also seen a number of ambitious gardeners use 1/2-inch solid drip hose to supply a Medusa's thicket of spaghetti tubing with spot spitters or emitters. And I've also watched these same gardeners curse in frustration: at the tangled mess of tubing, at how easily the spaghetti tubing pops off the drip hose, at how readily the emitters separate from the spaghetti and at how often the tubing is damaged while weeding. Some ended up, like Hercules burdened down by the manure of the Aegean stables, dragging large heaps of snarled spaghetti tubing to the local dump. As far as I'm concerned, spaghetti tubing is out of the question with annual plants because of their need for periodic care and seasonal cultivation.

Drip hose with punched-in emitters would have been the next logical choice for annuals, but I found out that the heat of summer and the sun's ultraviolet light soon cause the rigid plastic of the emitter to become brittle. Then, during routine tasks such as weeding, tilling or hoeing, the brittle barb snaps off at the base of the emitter. When this happens not only must the gardener find the hole, plug it with a goof plug and punch in a new emitter, but the end of the drip line must also be opened to flush out the broken tip of the barb. Snapped-off emitters under mulch can go undetected, and not only are large amounts of water wasted but the soil in that area also gets puddled and anaerobic. While certainly superior to the three versions of thin-walled soaker hoses, and to spaghetti with emitters, this still didn't seem to be the ultimate solution.

The best material didn't present itself to me until 1982, when I was tending a booth at an agricultural trade show. The booth next to mine featured something I'd never seen before, drip hose with the emitters preinstalled inside the hose. I immediately tried some out on rows of fruit trees and was pleased, but it was another four years, after lots of testing in various landscape settings, before I attempted using in-line tubing in a vegetable bed.

In-Line Emitter Tubing with Raised Vegetable Beds

I almost always design vegetable beds for my clients' edible landscapes as raised beds with 2- by 12-inch wooden sides. By a remarkable coincidence, this is exactly what Mr. and Mrs. Shrubbery have in the section of their landscape just beyond the lawn (**See Figure 59**). The wooden boxes solve a number of problems: There is a clear demarcation between the

edge of the path and the beginning of the bed. The wood helps keep the gravel or wood-chip mulch of the path from mixing with the soil in the bed. The raised soil allows for good drainage during our wet, rainy California winters, and a layer of 1/2-inch aviary wire, which looks like a smaller, sturdier version of chicken wire, can be added between the box and the ground to exclude destructive gophers and moles. Although the boxes typically are 3 feet wide and 8 to 10 feet long, the first time I tried in-line emitter tubing as a vegetable drip system was for a very small raised-bed design at the Preston Vineyards and Winery near Healdsburg, CA. My goal was to design a harnes,s or assembly, of in-line tubing inside the vegetable boxes which could easily be removed from the boxes by one person in order to cultivate the soil. Due to Sue Preston's design criteria (she needed only enough vegetables for daytime salads and snacks for the winery staff), the beds for this design were a petite 3 by 6 feet. The harnesses, composed of

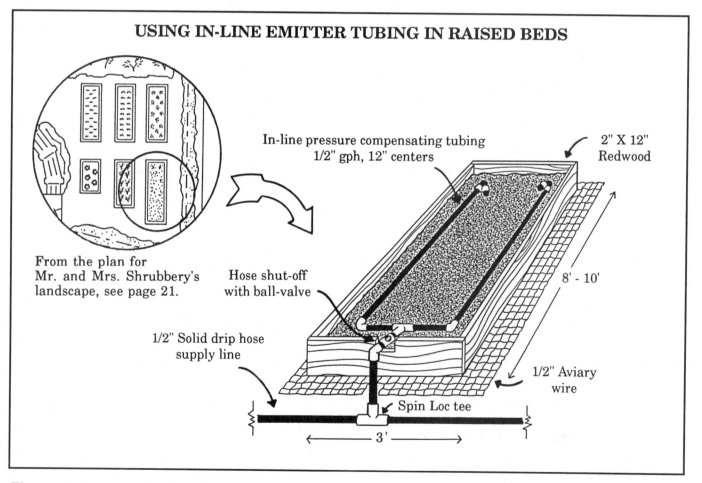

USING IN-LINE EMITTER TUBING IN RAISED BEDS

From the plan for Mr. and Mrs. Shrubbery's landscape, see page 21.

In-line pressure compensating tubing 1/2" gph, 12" centers

2" X 12" Redwood

Hose shut-off with ball-valve

1/2" Solid drip hose supply line

8' - 10'

1/2" Aviary wire

Spin Loc tee

3'

Figure 59 A convenient and simple way to water vegetables with drip irrigation was always a challenge until I started using this system. The hose shut-off ball-valve allows the gardener to turn off irrigation to beds without plants. The aviary wire is only required where gophers, moles or voles are a problem.

two lines of in-line emitter tubing, for each box were easily removed and reinstalled by the quick unthreading of a single hose swivel fitting in each bed.

The one drawback with drip irrigation for vegetables is that small-seeded varieties which are grown rather close together, such as carrots, arugula, beets and turnips, must be hand-watered until they show their first or second set of true leaves. The wet spot formed by drip irrigation on the soil's surface isn't usually wide enough to germinate small seeds broadcast over the entire area. Shortly after sprouting, however, these plants' taproots become deep enough to tap into the continuous wall-to-wall drip zone of moisture, 4 to 6 inches beneath the soil's surface. Larger, big-seeded plants can be sown near each emitter

and germinated solely with drip. Presoaking the larger seeds overnight in a diluton of water and seaweed powder will help ensure healthy germination.

Since I conceived the design for the Preston Vineyards, I have modified some of the parts. **Figure 60** shows my current configuration. A main line of 1/2-inch solid 600 or 700 series drip hose is laid in a trench in the ground along one end of all the boxes. At the center point of each box, the solid hose in the trench is cut with hand pruners and a Spin Loc tee is inserted with its "leg" facing up. Enough solid hose is added to come out of the ground and reach the top edge of the box, then the main line is flushed and the trench is backfilled. In the center of the box, a 1.5-inch-square notch is cut into the upper lip of the 2- by 12-inch board.

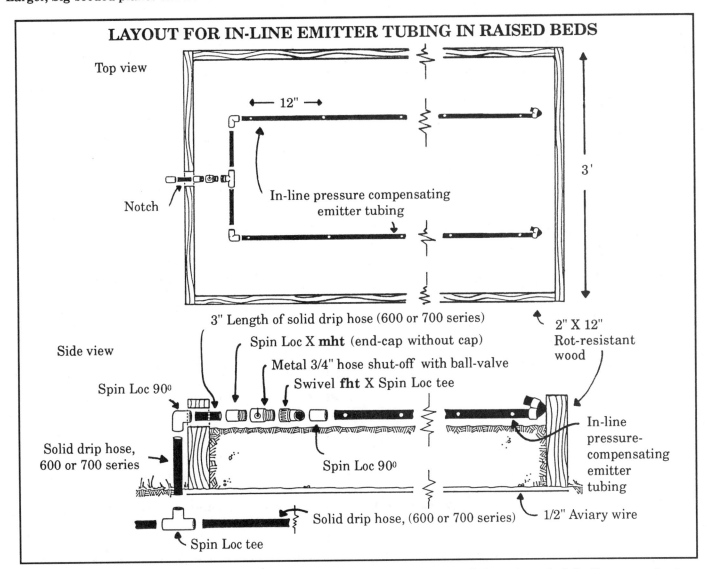

LAYOUT FOR IN-LINE EMITTER TUBING IN RAISED BEDS

Top view

12"

3'

Notch

In-line pressure compensating emitter tubing

3" Length of solid drip hose (600 or 700 series)

Spin Loc X **mht** (end-cap without cap)

Metal 3/4" hose shut-off with ball-valve

Swivel **fht** X Spin Loc tee

2" X 12" Rot-resistant wood

Side view

Spin Loc 90⁰

Solid drip hose, 600 or 700 series

Spin Loc 90⁰

In-line pressure-compensating emitter tubing

Solid drip hose, (600 or 700 series)

1/2" Aviary wire

Spin Loc tee

Figure 60 The most important part in this system for drip irrigating vegetable beds is the swivel fht X compression tee. With the twist-of-a-wrist, you can separate the in-line emitter tubing from the ball-valve, remove the tubing to an out-of-the-way place and cultivate without obstructions. The beds should be no wider than three feet to prevent lower back pain.

This permits the drip hardware to pass into the box without visually breaking the line of the end board. The following series of parts is assembled to allow for speedy and convenient removal of the drip irrigation harness from the box:

1—Add a Spin Loc elbow to the top of the solid drip-hose riser coming out of the main line. Trim the riser so that the elbow will pass its drip hose through the notched area.

2—Insert a 3-inch length of solid drip hose into the end of the elbow that is parallel to the ground.

3—To the end of the short piece of solid drip hose, add a Spin Loc X mht fitting. These are sometimes sold in a set with a fht cap as end-closures for drip lines. Just save the cap for that wonderful collection of parts kept in a large coffee can and labeled "Plumbing stuff I have no idea if I'll ever use but I'm too cheap to throw away."

4—To the mht, add the female end of a metal ball-valve hose shutoff. This will allow you to turn off certain beds if nothing is planted in them or remove the harness from one bed while other beds are being irrigated.

5—Some hose shutoff ball-valves have little holes built into each side of the valve. Use these holes to tap some small nails into the base of the wooden notch to secure the plumbing assembly, or use two screws with washers to hold the ball-valve in place. Be sure the male hose-threads of the hose shutoff extend *beyond* the wood, toward the inside of the box.

6—Next, add a Spin Loc swivel X compression tee to the mht of the shutoff valve. The "leg" of the tee has a swivel hose fitting, and the "top" of the tee has two Spin Loc fittings for in-line pressure-compensating emitter tubing.

Now you can use in-line pressure-compensating emitter tubing to finish the harness, which rests on top of the soil in the bed. I usually recommend the in-line tubing with 1/2-gph emitters on 12-inch intervals because it works on both sandy-loam and clay soils, depending upon how long you run the system. One- and 2-gph emitters, with their higher flow rates, work better in very sandy soils because the greater volume causes the water to spread more horizontally underground. This makes for a fatter carrot-like shape to the wet spot beneath the surface of the "bed," and distributes water more evenly to the roots. For a 3-foot-wide bed with a fairly loamy clay soil, you'll probably need only two lines of tubing running the length of the box. A 4-foot-wide bed may need just three lines of tubing. If you're growing a crop which is planted close together, such as garlic, beets, turnips or small European butter lettuces, you may want to have four lines per box to ensure adequate irrigation. It's best to put together a trial system first to see how the harness configuration works with your soil. In any case, to get the spacing (in inches) between each line across the width, divide the bed's width (in inches) by a number which is one larger than the number of lines you'll be using.

From the swivel tee, next to the shutoff valve, add the appropriate lengths of in-line tubing to reach the next fittings on the main header which runs the width of the bed. For the outer two lines, add Spin Loc elbows. For any lines other than the outside ones, add tees at the appropriate intervals. Always remember to leave these lines loose and *not* firmly staked, to protect the fittings as the tubing shrinks in the fall.

Flush the system. Cap the ends of each line in all the vegetable beds. Mulch to disguise and protect the tubing, and you're ready to grow some food.

Remember, your solid main line buried along one end of the boxes can supply no more than 240 gph. Determining the number of lateral lines running off one supply line can be done in two ways. First, simply use the Maximum Length of Tubing chart in the **Appendix** to determine the maximum number of feet for the type of emitter tubing you are using and make sure each lateral is equal to or less than the recommended length. If, for example, you're using in-line pressure-compensating tubing with 1/2-gph emitters on 12-inch spacings, make sure the total length of all the in-line tubing coming from one solid drip hose supply line to your vegetable boxes is less than 326 feet.

Sometimes the flow-rate capacity of the emitter tubing (from the Appendix) exceeds what the solid drip hose can supply. For example, the chart indicates that in-line pressure compensating tubing with 1/2-gph emitters on 24-inch centers can have a maximum length of 584 feet. But remember, the solid drip hose can supply only 240 gph. Technically, therefore, you must have three separate supply lines to service 584 feet of this tubing—two solid drip hose lines supplying 240 gph and another line to supply the remaining 52 gallons. In reality, there is enough of a fudge factor in these charts that you could divide up the 52 gallons among the other two lines and probably not notice a significant drop in the flow.

The other approach is to total the flow rates of all the emitters and make sure that number is less than 240 gph. If, for example, you had chosen to use 2-gph emitters on 12-inch centers, the maximum length would be 120 feet—240 gph divided by 2 gph = 120, times 1 (the interval of the emitters, in feet). If, in another example, you were using 2-gph emitters which you've punched in on 18-inch centers, the maximum footage would be 160 feet—240 gph divided by 2 gph = 120, times 1.5 (the emitter interval in feet) = 180 feet. In this example, had you planned to use 1/2-gph emitters on 18-inch centers, the maximum length of tubing would be 320 feet—240 gph divided by .5 gph times 1.5 feet.

Drip for Vegetable Beds without Wooden Boxes

Not everyone wants to go to the expense of wooden vegetable beds. The simplest and cheapest beds are merely dug in place, and often the only way to tell a pathway from a bed is by the vegetables growing in one of them. One considerable advantage to unboxed beds is that their shape can form any part of a curve or circle—an aesthetically pleasing alternative to boxes.

There is a simple way to connect the same type of harness in a ground-level bed with a buried main supply line. Without the gopher-retarding wire bottom of a box to deal with, you'll be digging much deeper into the native soil as you cultivate. To protect the main water supply line which goes from bed to bed, use schedule 40 PVC pipe instead of solid drip hose and place it in a trench 18 inches deep, well out of reach of seasonal cultivation. A 3/4-inch main supply line less than 200 feet long, will supply 480 gph. A one-inch main line can deliver up to 780 gph.

I don't recommend rigid PVC pipe for the risers from the buried main line to the beginning of a bed harness. A wheelbarrow or even a clumsy foot bumping into a rigid PVC riser, even if it's clamped to a post, can transmit enough of a shock to break the schedule 40 tee fitting on the main supply line. There is a flexible PVC hose called IPS Flex-PVC, which can be glued into regular schedule 40 PVC fittings. If you bump into the IPS Flex-PVC hose, it's limber enough to bend without snapping or cracking the schedule 40 tee below ground. With IPS Flex-PVC hose, the 3/4-inch hose can pass 480 gph and the 1-inch hose, 780 gph. Be sure the total footage of each harness falls within this range.

Now, refer to **Figure 61**. Wherever you want a riser along the main supply line, cut the pipe with a hacksaw and use regular PVC glue (keeping in mind the likelihood of PMLP #7 occurring) to affix the correct-sized s X s X s (all slip or glued) tee to the main line with its open "leg" facing straight up. (If you're using a 1-inch main supply line because it must serve a lot of beds, use a 1-inch s X 3/4-inch s X 1-inch s tee.) Use a special clear PVC glue such as Weld-On™ #795, which is made to cement 3/4-inch IPS Flex-PVC hose to schedule 40 fittings, to attach enough hose to reach just above the soil's surface. Glue a 3/4-inch s X s schedule 40 elbow to the top of the IPS Flex-PVC hose with the open end facing the beginning of the bed. (Make sure to use the correct glue.) To one open, or slip, end of the elbow, cement a short, 3- to 4-inch length of regular 3/4-inch schedule 40 PVC pipe, using ordinary PVC glue, and then glue a 3/4-inch s X mht transition fitting to the end of the PVC pipe. To this male hose-thread you can add the ball-valve hose shutoff and the rest of the bed's harness, as described above.

Row Crops and Drip Irrigation

If you prefer to garden in the time-honored fashion called row cropping, with wide paths between rows of vegetables, you can still use drip irrigation. Row cropping is well suited to big patches of corn, long rows of tomatoes for canning, trellises of cukes for pickling and lengthy rows of dry-shell beans. In some ways, row cropping is the easiest way to use in-line drip tubing because no harnesses are required.

You can use a main line of either solid drip hose or schedule 40 PVC pipe and plumb off of the main line as described above. The only difference is that you'll want to have each row's drip hose attached separately to the main line by its own riser with a fht swivel fitting. This makes it easier to disconnect the long lines of drip tubing and drag each one off to the side before

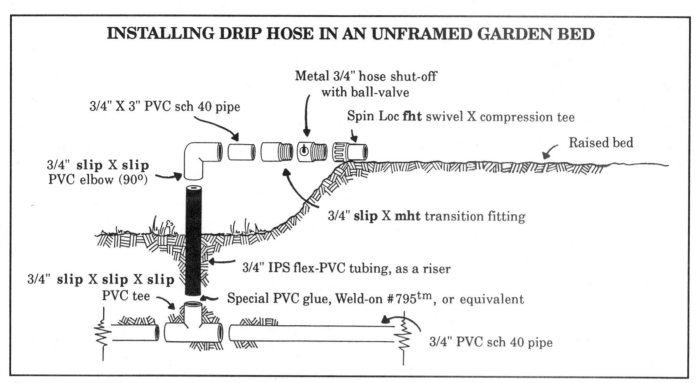

INSTALLING DRIP HOSE IN AN UNFRAMED GARDEN BED

3/4" X 3" PVC sch 40 pipe

Metal 3/4" hose shut-off with ball-valve

Spin Loc **fht** swivel X compression tee

Raised bed

3/4" **slip** X **slip** PVC elbow (90°)

3/4" **slip** X **mht** transition fitting

3/4" IPS flex-PVC tubing, as a riser

3/4" **slip** X **slip** X **slip** PVC tee

Special PVC glue, Weld-on #795^{tm}, or equivalent

3/4" PVC sch 40 pipe

Figure 61 If you don't want to spend a lot of money on wooden beds for your vegetables, drip irrigation can still be used with row crops or borderless raised beds. Use flexible IPS tubing to help prevent damage from a wayward foot, hoe or wheelbarrow. The deeper you bury the supply line, the safer the PVC tee is from damage when the IPS tubing is knocked around. The configuration of the in-line tubing depends on your planting pattern.

rototilling the entire plot. When hoeing, simply flop the tubing over to an already cultivated portion of the row, hoe where the tubing was and flop it back.

Vining winter squash and pumpkins have such an extensive root system **(See Figure 62)** that it's better to drip-irrigate more as you would with trees than with row crops. Use a large spiral of in-line tubing to water as much of the root zone as possible.

How Often and How Long to Water Your Vegetable Beds

Even more than with flowers, trees and shrubs, I suggest experimenting with daily applications of tiny amounts of water on vegetables. Unless your water supply is extremely limited, judicious daily waterings will produce the greatest yields, although with some crops, like tomatoes, to produce a more richly flavored fruit you may want to reduce the amount of water as the plants begin bearing. Too much irrigation will plump up the 'maters with water; this dilutes the flavor and makes it more difficult to cook them down into sauce or paste.

Happy harvesting!

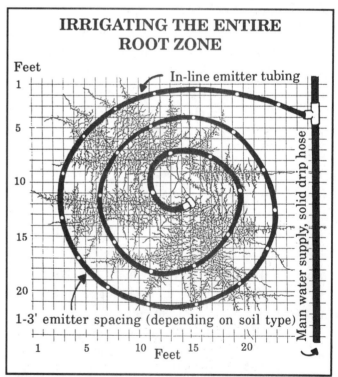

IRRIGATING THE ENTIRE ROOT ZONE

Feet

In-line emitter tubing

Main water supply, solid drip hose

1-3' emitter spacing (depending on soil type)

Feet

Figure 62 The radius of an 11-week-old winter squash's roots can be over 12 feet in all directions. Irrigating the entire root zone with plenty of emitters will ensure the best growth and highest yields. Be sure the emitters don't water the crown of the root system at the stem.

9 Keeping Your Drip Together

All your investment in time, sweat, profanity and money will be for nothing if you don't perform a little routine maintenance on your new drip irrigation system. Unlike a regular garden hose, drip irrigation requires some periodic attention. This slight extra effort will be repaid many times over.

Flush Your Troubles Away

The first three rules of drip maintenance are flush, flush and flush. As mentioned earlier, the drip irrigation hose should be flushed when it's first laid out in the garden, again after any emitters or smaller-diameter tubing are punched into the 1/2-inch hosing and before the ends of each line are capped. Once your drip irrigation is installed, the more you open the ends of the lines and flush the 1/2-inch hosing, the safer you'll be from conjuring up some new Pop Murphy's Law of Plumbing. Each line in the system should be flushed at the beginning of every growing season and *at least* once during the season. As with the original installation, open all the lines, run the water for several minutes, begin closing the lines at the faucet or the highest lines and work your way progressively away from the faucet and downhill.

Even with a quality Y-filter, some sediment and crud may slip through the filter and into the irrigation lines. At one time, I thought dripology researchers had licked this problem of renegade filth and was led to experiment with a gizmo called a "flushing end cap." The concept sounded great. Each time the system was turned on, the flushing end cap would remain open for a short period of time to allow some water, and its sedimental contraband, to escape. As the pressure in the line built up, over a minute or two, the flushing end cap would shut off and the emitters could get to work irrigating the landscape. In the real world, things went awry. First, the little springs didn't always do their job and the flushing end caps kept flushing the entire time the system was on. Second, people with well water (unchlorinated) occasionally found bits of slime clogging the end cap and preventing proper sealing under pressure. And finally, an open end cap was an open invitation to slugs, earwigs and pillbugs to set up house.

These pesty critters can cause far more havoc than any amount of sediment suspended in the water. So my flushing end caps are safely stored in the box marked "Sounded Like a Great Idea, but Doesn't Work."

Fortunately, dripology researchers continued to experiment with and redesign flushing end caps. One newer model which looks very promising is called a line flushing valve. **(See Figure 63.)** This particular automatic flushing device utilizes, in part, the same tortuous path technology used with in-line emitter tubing. Before the drip system is turned on, several diaphragms lie limp with four small open ports to allow for the flushing of accumulated sediment. The small outlets make it difficult for bugs to enter but easy for water to exit. Any insect which might enter the body of the valve via the ports would have to crawl through the entire labyrinth of the valve before entering the drip hose. Each time the drip system is turned on, about a gallon or more of water is allowed to pass through the valve's tortuous path, flushing out any minute debris. Then the diaphragms inflate and seal the flushing valve shut, and the drip system functions normally, at full pressure. After the drip system is turned off, the line flushing valve allows water to drain out of the lines, again removing any wayward sediment. The lack of standing water in the lines prevents the buildup of algae scums and iron-water bacterial slimes—both of which can clog emitters, especially with unchlorinated well water. The line flushing valve is best placed inside a purchased or homemade valve box over a pocket of gravel. If you use the version with the 1/2-inch mipt fitting, add a fipt fitting to the end so the line flushing valve can be easily removed for a more lengthy manual flushing. With the barbed fitting, install a tee with two barbs and, on the leg of the tee with a mht, add a threaded fht end cap. This allows for flushing without removing the valve's barb from the drip hose. I've just begun to try this line flushing valve in my garden, but early experience looks encouraging.

A Filthy Filter Is a Terrible Waste

A drip system's emitters can clog *much* faster than kidney stones forming in a thirsty cheese-eater.

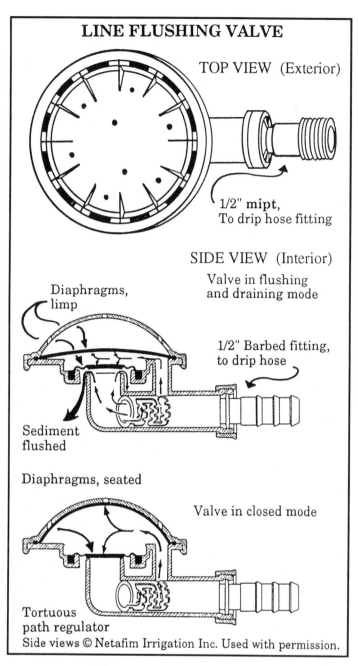

LINE FLUSHING VALVE

TOP VIEW (Exterior)

1/2" mipt,
To drip hose fitting

SIDE VIEW (Interior)

Valve in flushing
and draining mode

Diaphragms,
limp

1/2" Barbed fitting,
to drip hose

Sediment
flushed

Diaphragms, seated

Valve in closed mode

Tortuous
path regulator

Side views © Netafim Irrigation Inc. Used with permission.

Figure 63 At the beginning and end of each irrigation cycle, a line flushing valve automatically discharges any sediment which has accumulated in the drip hose. Install with the dome facing up.

While the results are nowhere near as painful, the end result is similar—clogged tubing where water should easily flow. A properly designed drip irrigation system has a filter to protect the emitters, Laser tubing or porous tubing from getting blocked with sand, silt or sediment. The filter may stop all the filth, but if it is left uncleaned, the flow of water to your emitters is greatly reduced. A clean filter insures clog-free emitters *and* adequate flow for even distribution of water.

The Y-valve with a ball-valve, as recommended earlier, makes it easy to flush the filter cartridge regularly. You should flush each Y-filter *at least* once a month, and gardeners with well water may find a *weekly* flushing is required.

If you install an electronic irrigation clock to control 24VAC solenoid valves (which will be discussed in the next chapter), there is a sneaky way to flush your filter automatically. At the end of the filter's flush ball-valve, install an irrigation solenoid. **(See Figure 64.)** Leave the filter's flushing ball-valve in the *open* position. Wire the solenoid to your electronic irrigation clock, but to a separate "program" from the irrigation lines (again, soon to be discussed in detail). Then you can set the controller to turn on that program once a week for one or two minutes to flush the filter automatically.

No matter how much you flush your Y-filter, the water streaming by will not strip everything off the metal screen, especially algae and other unrecognizable slimy stuff. So, at the beginning of *every* irrigation season and *at least* once each summer, take the Y-filter apart, remove the metal screen cylinder and scrub well with an old toothbrush and a strong solution of bleach.

Give Your System Some Acid

Despite all these precautions and tidy compulsions, some buildup may occur on the emitters. This is especially true where the water supply is high in minerals, such as soluble iron or calcium. No filter can clean these soluble minerals out of the water while they are in solution. As the minerals tumble along the dark labyrinth of your drip system, the agitation and contact with oxygen will precipitate some of them—especially iron—into solids. These particles can build up on the interior surfaces of the emitter, as in hardening of the arteries. Sometimes the soluble minerals don't get enough exposure to air to precipitate until they reach the opening of the emitter, in which case you'll see a buildup of minerals around the orifice of each emitter.

In either case, there is, so to speak, a solution. Once a year, or as needed, you can dissolve mineral buildup by pumping a dilute solution of acid through your drip irrigation system. Sounds gruesome, but it's

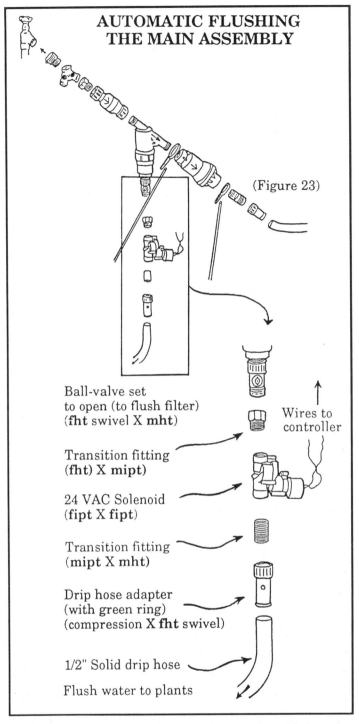

AUTOMATIC FLUSHING THE MAIN ASSEMBLY

(Figure 23)

Ball-valve set
to open (to flush filter)
(fht swivel X mht)

Wires to
controller

Transition fitting
(fht) X mipt)

24 VAC Solenoid
(fipt X fipt)

Transition fitting
(mipt X mht)

Drip hose adapter
(with green ring)
(compression X fht swivel)

1/2" Solid drip hose

Flush water to plants

Figure 64 An electric solenoid valve can be added to the flushing end of the Y-filter to automatically clean the screen via one of the digital controller's separate programs. Be sure to check this setup after installation to make sure the controller is set to the proper time and program.

not too difficult or risky. First, you'll need to plumb into your system a special gizmo called an injector, or proportioner. This device draws (suctions) a concentrated solution from a pail or bucket while the drip system is running and proportions a dilution of, in this case, acid into the stream of water in the drip hose. (This device can also be used, with thorough filtering, to inject a dilution of liquid fertilizer into the drip system for automatic feeding.)

You must purchase a low-volume type of injector/proportioner. These are available from a number of the mail-order companies in the Suppliers and Resources listing in the Appendix. Be sure to use a low-volume proportioner. Many proportioners, especially the inexpensive types, require more flow to suction the concentrated solution from the bucket than a drip irrigation system can provide. The inexpensive proportioners sold at many hardware stores are designed to function with the flow of a regular garden hose and oscillating sprinkler (5 gpm) and will not work at the low flow rates of a typical drip irrigation system. Each proportioner requires a minimum flow rate to begin the suction of liquid concentrate into the system. Check the box or instructions to make sure the proportioner will operate at the flow rate, in gpm or gph, of your drip system. For example, one model of a proportioner called the Dosatron™ works at a flow rate between 2.64 and 660 gph, while another Dosatron™ model functions between 300 and 6000 gph, and the Rainjet™ version of a fertilizer injector (as drawn in **Figure 65**) operates between 30 and 300 gph.

Total up the number of emitters in your system and divide by the flow rate to get the flow in gph. For example, a series of lines of in-line tubing with 1/2-gph emitters every 12 inches totals 224 feet. This equals 224 emitters times 1/2 gph, or 112 gph. Thus, two of the proportioners mentioned above will work. But if you had a series of hanging plants with a total of 15 1/2-gph emitters, for a flow of 7.5 gph, these proportioners wouldn't have enough flow rate to function.

The proportioner must be plumbed in between the backflow preventer(s) and the filter and pressure regulator. **(See Figure 65.)** Placing the proportioner before the filter (upstream) allows the filter's screen to catch any wayward dirt, sediment or undissolved chemicals or fertilizers. The fittings required to plumb the proportioner into the main assembly will vary with the model.

To flush your drip system, you simply mix up a solution of any one of the emitter cleaners on the

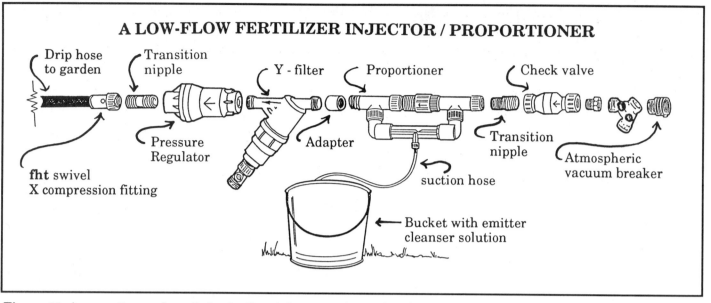

Figure 65 A proportioner, also called a fertilizer injector, can be used to clean your emitters by pumping a dilute acid solution through the drip irrigation system. Make sure the proportioner is installed between the backflow preventer(s) and the filter.

market, pour the solution into the bucket and turn on the drip system. Follow the manufacturer's instructions for both the dilution rate and the length of flushing. This is best done during the dormant season or during a heavy rain to mitigate the effect of the acids on the soil. Commercial orchardists in the Central Valley of California have been routinely flushing their drip systems for years with no deleterious effects.

One emitter-cleansing product, Drip-A-Tron Cleaner™, purportedly cleans and lubricates the drip irrigation system with a solution of 5% hydrochloric acid, 10% phosphoric acid and 2% copper sulfate, among other chemicals. This material doesn't qualify for use on a California certified organic farm because some of the chemicals are derived from fossil fuels. For all but the certified organic farmers, these chemicals are represented by the manufacturer as biodegradable and nonpolluting.

Winterize Your Drip System

A proper drip irrigation system is too valuable to leave neglected like an abandoned garden hose at each winter's nippy return. Shut off your system's water supply at a point inside the house where the valve is protected from freezing. Loosen all unions at each main assembly to drain all water out of the pipes, filters, pressure regulators and check-valves. Make sure you don't lose the O-ring which seals each union. Be sure to

open the flush ball-valve on the Y-filter so the filter chamber doesn't crack. Better yet, bring the filter or the entire main assembly indoors. Also bring inside any battery-powered controllers which you have attached to faucets.

The drip irrigation hose can stay in place in the landscape, but as much water as possible should be drained out. Open the end-closures on all lines, let all water drain out and close. Use threaded end caps because the figure-8 end closures will quickly wear out the kinked end of the hose. A deep mulch will help protect the plastic tubing in cold winter climates. You don't want anyone stepping on cold plastic tubing, but since many drip systems are primarily located in perennial plantings, the risk of trampling the tubing is low. Some people do lift out the entire assembly of header and lateral lines, roll up the tubing and store it in the garage for the winter. Porous pipe and in-line emitter tubing make this a lot easier than hose with emitters punched in or hose with, God forbid, spaghetti tubing with emitters.

Nearly all sprinkler systems are designed with automatic drain valves. These simple devices stay open when the system is first turned on, flush water until the pressure builds, usually to 2.5 psi, then close. When the sprinkler system is turned off, the drain valve automatically opens and water remaining in the lines drains out, usually into a small gravel sump. Like the ill-fated flushing end cap, the drain valve has its drawbacks. First, a considerable amount of water is flushed out every time the system is turned on throughout the summer, merely to drain the lines the last time you shut off the system for the winter. Second, there is a possibility that the valve won't seal properly, and some dirt particles may be sucked back into the drip lines to clog the emitters.

Installing a manual drain-down valve is much

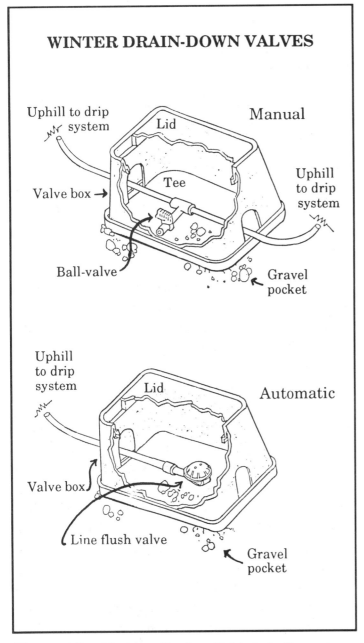

WINTER DRAIN-DOWN VALVES

Figure 66 In cold-winter climates, all drip irrigation systems must be drained in order to protect your investment. Either install a manual ball-valve which is opened each fall or a line flushing valve which automatically drains the lines each time the system shuts off or is turned off.

safer and conserves far more water. If possible, lay out your drip irrigation lines so they can all be drained from the low ends. Otherwise, install your drip lines so they naturally drain to various low spots, install an access box at the lowest spot in each line and plumb in a tee with a simple ball-valve. Each fall, the plastic drip irrigation ball-valve should be opened to fully drain the line; see Figure 66.

Lonnie Zamora, a rooftop-garden irrigation designer in New York City, has been using in-line emitter tubing for over four years on most of his jobs, even in the flower boxes on the 27th floor of a Manhattan penthouse. To winterize his client's drip irrigation systems each fall, he (1) drains the pressure regulator; (2) closes the faucet at the home's wall; (3) attaches a 110VAC air compressor with the necessary fittings to the faucet at the beginning of the drip system; (4) sets the clock controller to run each line or zone for five to seven minutes; (5) runs the air compressor pumps at 25–40 psi to blow out all the lines; (6) manually drains all lines on multilevel, terraced gardens and, finally, (7) opens the Y-filter's ball-valve to drain the filter cartridge chamber.

Tidy Tips for Your Drip System

Every piece of rigid PVC pipe which shows aboveground should have been painted with a quality exterior latex paint right after installation to protect the pipe from the degrading effects of sunlight. If you haven't done this, do it! Besides protecting the pipe, you'll help disguise it. Choose a color which matches your house paint or an earthy brown which easily blends into the background. Occasionally, some of the paint may flake off the pipe. Repaint as needed.

Keeping Your Drip Together

No matter how much time you spent planning and installing your drip irrigation system, you'll think of new ways to improve and change it, or there'll be the occasional repair. For such occasions, the drip irrigation emergency repair kit comes in handy. (See Figure 67.) I got this idea from Sidney Ocean, head gardener at the Esalen Institute in Big Sur, CA. She has all her small miscellaneous drip irrigation parts neatly arranged in a sewing box. The box, like some fishing tackle boxes, has

STORING MISCELLANEOUS DRIP PARTS

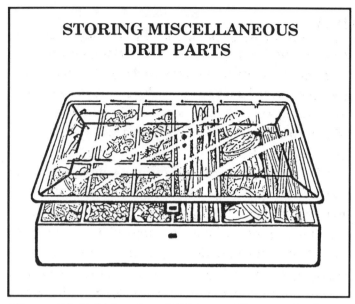

Figure 67 Multichambered fishing tackle or sewing boxes make great places to store miscellaneous parts. Such boxes, full of emitters, widgets and drip paraphernalia, make good repair kits for the inevitable occurrence of many of Pop Murphy's Laws of Plumbing.

lots of small divided chambers and a clear lid so you can see what is stored in each little cubicle. No longer will you have to rummage among various paper bags or coffee cans to find the parts you want to schlep out into the landscape—only to find, once you start, that PMLP #1 has insured you're missing a part. Goof plugs, various types of emitters, transfer barbs, mister heads, and hose punches can all be neatly stored and easily located when repairing or improving your drip system.

10 Controlling Your Drip Irrigation

So now you've got a new drip system. Perhaps, due to leaks, you've sworn off all plastic-made products, or the pipes burst, which helped precipitate your divorce, and all your plants have died and your former father-in-law is coming to live with you! Most probably, though, you've fully triumphed over Pop Murphy's Laws of Plumbing and have a discreetly hidden, efficient and properly designed drip irrigation system.

But don't award yourself the Degree in Dripology found in the **Appendix** yet. You must still master the Zen koan of drip irrigation—What is the sound of one valve turning? To control your drip system is to realize the art of refined simplicity. To fail means that your drip system will control you, or at least a lot of your free time as you scurry about troubleshooting electric valves and digital controllers.

How to control this far-flung petrochemical octopus? The choices range from the merely manual to the electronically ingenious—or devious, if you don't like computerized gizmos. I will explain ways to control your drip system beginning with the easiest and ending with the most complicated. The choice is yours. But, at this point, my lawyers advise me to inform you that I cannot be held liable for the bizarre things which can go wrong once you cross over into the realm of microchip irrigation gadgetry.

A Twist of the Wrist

Throughout this book, except for **Figures 45 and 64**, I've discussed and presented drip irrigation systems with manual valves: sometimes faucets, at other times ball-valves. There are three very good reasons for using such valves: simplicity, simplicity and simplicity. Manual valves are what I affectionately refer to as "Italian plumbing," meaning a solution which is not overly engineered and is inexpensive, easy to operate, plain and elegant in its own way. In other words, "Why-a you make-a life more complicated than it already is? Heh?"

Most people don't need any instruction in turning on a manual valve. But the trick to manual valves is remembering to turn them off. I should know; I gave some of my plantings some very deep waterings

(actually horizontal, since there's a clay pan less than 18 inches from the surface) before I relied upon a simple device to remind me to shut off the system. I use what I call a cake timer (more often called a kitchen timer), one of those old-fashioned spring-wound timers which allow you to set the knob anywhere from several minutes to one hour. With a simple twist of the wrist, you can set the timer to the length of irrigation you've calculated earlier based upon the season's ET rate. The cake timer reminds you with a loud ring that it's time to ramble outside and turn off the drip system. Beware: some models don't really begin to tick until after the knob is turned past the five-minute mark. With such timers, shorter daily run times can easily be controlled while gardening or during your morning coffee or tea.

The only drawback to the cake timer, besides your having to remember to set it in the first place, is the infernal tick . . . tick . . . ticking. This straightforward mechanical device does make a loud, irritating sound. I leave mine outside my writing studio's door, down the hallway far enough so the ticking isn't very noticeable but the distinctive clang is easily heard.

Fancy kitchenware catalogs and stores now sell digital kitchen timers. Although they use batteries and are harder to set, digital kitchen timers offer the distinct advantage of no obnoxious ticking. Why I don't have one only my therapist knows for sure.

Manual-On, Automatic-Off Controllers

The next gentle step up in complexity involves those timers which you turn on but a gizmo turns off. Such a device gets rid of one-half of the equation of irrigation glitches: short-term memory loss. Gone is the task of remembering to set the kitchen timer after you've turned on the system. Gone is the tedium of setting the digital kitchen timer.

But are these devices reliable? Alas, certain models are not really fail-safe. There are two very different versions of these ticking, self-closing timers, the commercial grade and the home gardener model. There is one popular European model marketed through many mail-order catalogs which is known by its distinctive colors. The gauge on the knob is calibrated in minutes and ranges from five minutes to two hours. I know many people who have had good results with this cake timer version. But I do know of a large commercial garden project which had purchased several dozen timers. The owners of the garden found that at least 30% of the timers broke at some time. Even though the company was very courteous and prompt in replacing the defective units, the hassle and time required to switch the broken models with new ones was considerable. If Pop Murphy is sleeping, you'll be very lucky and the timer will break while shut off. Unfortunately, this type of timer is prone to break while on—leading to an undetected trickle of uncontrolled water. Such timers are sold for a few dollars short of $30.

The next step up in quality costs an additional $30. You can purchase a rugged, sturdy landscape timer called the Bermadon Automatic Metering Valve™, sold by the Bermad Company for about $60. **(See Figure 68**.) The valve comes in three models, each with a different minimum flow rate and capacity. The lowest-flow valve operates between 30 and 300 gph and between 7 and 85 psi. The numbers on the knob are calibrated at 100 gallons per notch. To irrigate, you simply twist to the total number of gallons desired and relax—the timer does the rest, regardless of fluctuations in pressure. The mipt threads on both ends are 1 inch and must be stepped up and down to the more common 3/4- or 1/2-inch garden drip irrigation fittings. Owen Dell of Santa Barbara, CA, one of the premier landscape designers in southern California, has long been dissatisfied with electronic controllers. Foremost, according to Dell, "the client doesn't grasp how to reprogram the controller and never adjusts the times from season to season. Plus, the controller is dumb—plants don't need water just because it's Tuesday and it's 6 a.m. Water need is a response to environmental factors, not just time." Dell is so impressed by the Bermad controller that he has almost entirely given up on electronic timers, except for use in turf areas. He also favors the Bermads because, in his words, "the homeowner maintains some active involvement with the landscape, can adjust the amount of water on a daily or weekly basis according to the

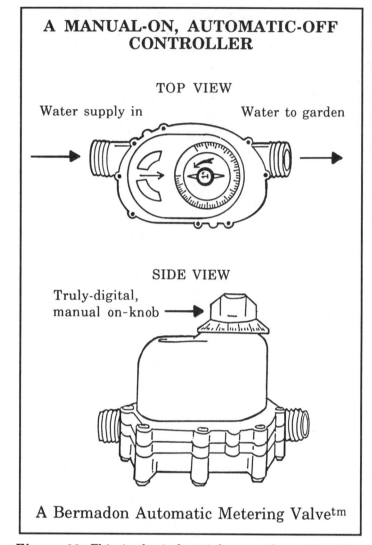

A MANUAL-ON, AUTOMATIC-OFF CONTROLLER

TOP VIEW

Water supply in   Water to garden

SIDE VIEW

Truly-digital, manual on-knob →

A Bermadon Automatic Metering Valve™

Figure 68 This is the industrial strength version of a manual-on, automatic-off irigation valve. The valve is set to a predetermined amount of water and it measures the flow regardless of fluctuations in pressure. One of the truly "digital" controllers.

weather and can watch for any malfunctions or broken hoses." The assembly in **Figure 69** shows a typical multistation setup as designed by Dell for his clients. Note that all pipes to and from the controllers are made of galvanized metal, for sturdiness.

Except for people who travel frequently, this system has a basic appeal. It is simple, sturdy, accessible and easy to use, and it keeps the gardener tuned in to the garden's watering needs. It is also now my system of choice because of my unsettling experiences with electronic controllers.

Computerized, Digital Irrigation Controllers

Another reason I favor manual controllers and

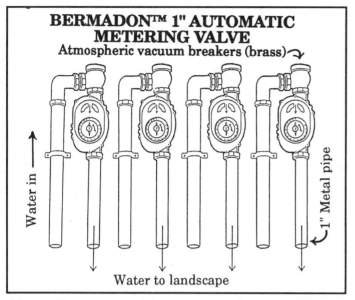

BERMADON™ 1" AUTOMATIC METERING VALVE
Atmospheric vacuum breakers (brass) ↘

Water in →

1" Metal pipe ↙

Water to landscape

Figure 69 An array of five Bermadon Automatic Metering Valves. This sturdy, Manually-controlled system utilizes metal pipe, heavy-duty "digital" valves and brass atmosperic vacuum breakers. The gardener can adjust the flow rate on a daily basis to accommodate subtle climate changes.

valves is that I really hate digital appliances, especially automatic drip irrigation controllers. I have mastered my microwave oven, because my very survival is at stake, but don't ask me how to program a VCR. Yes, I have managed to learn to program several makes of digital-readout automatic irrigation controllers, but I think of the process as a living purgatory at best. I remind you of PMLP #12; and I quote: "Pop Murphy was temporarily diverted from plumbing devices with the start-up of the Star Wars defense program. But now that he has trained the Pentagon in all his best management techniques, he has returned to designing the all-digital, all-confusing, microchipped, maxi-priced irrigation-controlling device."

The Faucet-Mounted Microchip Controllers

Many gardeners first experiment with automatic electronic controllers by purchasing one of the several models of stand-alone, battery-operated controllers which attach to the garden faucet. Some models use a 9VDC battery; others use AA batteries. In most cases, it is better to attach the controller to the faucet after the atmospheric vacuum breaker and the rest of the main assembly. It's especially important to use two legs to support the end of the main assembly by triangulation so the weight of the assembly doesn't stress the mht outlet fitting on the controller. Each

controller is digitally controlled and must be programmed to turn the water on and off for a specific length of time, usually referred to as the "run" time. First, you set the current time on a clock, then you enter the day(s) and duration of irrigation into the microchip. Various battery-powered controller models range in cost from $45 to $60.

There can be three problems with these controllers. First, although the language may be more readable than with many multistation controllers, in programming the device one often feels "slow" when trying to comprehend the owner's manual. Second, these controllers attach to one faucet and control only what is attached after the main assembly, for whatever run time you've programmed. In other words, you can't irrigate hanging pots, which require a very short daily run time, with the same controller used to water some fruit trees once a week for a long period of time. However, some models do allow you to program three different start times per day—which is also useful for short run times for sprinklers on a slope, to avoid runoff. Finally, the reliability of these gizmos may not be as good as their visibility in mail-order catalogs or hardware stores would suggest. I have one of those bicolored European models set slightly below ground level. It has been operating fine for four years, even with periodic winter flooding. Yet a friend, with an ambitious vegetable truck mini-farm, has purchased dozens and had up to 50% of them fail. Again, the company was very efficient and cordial with regard to replacement. But this didn't take care of wilting plants and the hassle of returning the defective units. I once used a model from a different manufacturer for a client, and the unit failed in the "on" position—leaving a considerable water bill legacy during the fourth year of the recent California drought! Yet another example of Pop Murphy's insidious influence. There are many new models on the market. Don't be bold and try out a new model without someone's recommendation. It is essential that you talk with other gardeners and get favorable testimonials about the model you're considering purchasing.

Although some models are designed to withstand freezing, the prudent gardener will protect his or her investment by bringing in these controllers when fall's cold breath begins to nip at the blushing leaves.

Multistation Digital Controllers

Now we can experience true gizmophobia. (See the **Glossary** in the **Appendix**.) Nothing makes me feel dumber than trying to decipher the instruction booklet of a new model of digital-readout computerized controller. Still, if you're going to automate the drip lines in your yard, I recommend your getting one of these multistation timers or controllers instead of a handful of stand-alone battery-powered units.

Foremost, choose the controller that is easiest for you to program. Sometimes the instructions are worse than those for a VCR. Simple programming is important in reducing your ulcer coefficient, in easily changing the duration of irrigation to reflect the seasonal changes in the ET rate and in reprogramming if the power goes out and the electronic instructions are lost to the wind.

Multistation timers or controllers can turn on a number of different drip irrigation lines by activating a 24VAC solenoid valve. Usually, a four-station timer, even with the cost of solenoids, wire and fittings, is less expensive than four separate stand-alone units. The least expensive models must be mounted indoors in a protected spot and near a 110VAC outlet. Wires can run as far as 1500 feet, depending on the wire's thickness or gauge, from the controller out to wherever the solenoid valves are located. Although the solenoid valves can be put anywhere in the landscape, most people cluster them together and place them as close to the controller as possible.

As with stand-alone controllers, electronic multistation controllers must first have the current time entered into the chip's memory. Once the correct time is displayed on the digital-readout panel, you will program the present day of the week, the length of irrigation, the days of the week for irrigation and/or the number of times an irrigation line will be turned on each day.

Each computerized controller is rated as being able to control a specific number of "stations." Each station equals one solenoid valve. Usually, there is a strip across the bottom of the back of the controller where you screw down the station wire from each solenoid. All the common, or ground, wires from the solenoids are combined to one white-coded ground wire and screwed to the common terminal on the strip. The minimal electronic controller is usually a four-station "clock" (landscaping jargon for a controller), which is able to turn on and off four different solenoid-activated lines or stations. Always buy a controller with at least a few more stations than you expect to use, as most gardeners find more ways to accurately control their drip irrigation systems as they refine their grasp of dripology or expand their plantings. Almost every landscape I have designed takes at least an eight-station controller as the watering needs are zoned, or grouped, fairly precisely.

Each controller allows you to schedule one or more "programs," those days of the week during which water comes on and the start and stop times for irrigation on those days. Less expensive controllers may have only one option, or one program. Other inexpensive models have two separate programs but four or six stations. In the case of a six-station clock, the six different posts for the solenoid's station wires can be arbitrarily divided up between the two programs. As an example, one of the controllers I've used, the Hardie Hydro Rain™ HR 6100, has a Program A and a Program C (they never explain what happened to B) and these two programs can be divided among six stations. For example, Program A could run four lawn sprinkler systems every day for short periods and Program C could be set to irrigate stations five and six for a long period of time, once a week. To encode the controller, you set the current time, punch in the current day, choose the days you want the program to come on (from one to seven or by intervals, such as every third day), punch in the length of the irrigation (the station or run time) and set the program's start time. Then all of the four solenoids wired to Program A come on in numerical sequence. With most, if not all, models, the controller cycles through each station during each program, one after another. Usually, the second station starts almost immediately after the first finishes; they can't overlap. For example, if lines one through four are set to start at 6 a.m. and run for 15 minutes each, they will stay on until 7 a.m. The other two stations, Program C, can be set to start at 8 a.m., run for two hours each and shut off at noon. Make sure you don't overlap the start times, as a larger supply pipe may be required to service the simultaneous flow of all stations. With sequential start times, the supply pipe doesn't have to be as large since all of the stations won't be on at the same time.

Some models allow each program to be set to start several or more times, with as many as 11 start times during the same day. This option is well suited to short irrigation times with lawns or misters in greenhouses but usually is not needed with drip irrigation. More advanced, and costly, controllers allow each station to be programmed independently from all the other stations. This allows for the easiest and most specific programming, but at considerable expense: one eight-station version costs over $500!

The electronic timer can control the lines for a range of times. Some can be programmed for zero to 99 minutes; others can be programmed for nearly six hours. Be sure you get a model which matches or exceeds the maximum run time you anticipate. Paying extra for a controller which has a longer run time will give you more flexibility and adaptability in the future. If you plan to irrigate container plants or hanging plants, double-check to make sure the clock can be set for times as short as one minute.

Some controllers are set up for a seven-day week. This makes programming awkward if you're planning to have the system come on every other day, because of the odd number of days in a week. Other controllers are programmed on a 14-day basis so you can easily set the program to alternate-day irrigation. Another option includes models which can be set independently of the days of the week to come on at any interval—every other day, every third day, every fourth day, etc.

Virtually all clocks have a manual override option. This allows you to turn on any station manually for extra spot waterings or to inspect the lines. The difference from model to model is the ease with which you can start a manual cycle. Make sure the instructions for manual irrigation are printed inside the door of the controller or on the control panel.

As with a computer, if the power to the controller is shut off, the microchip with the program you've set loses its memory. Many controllers utilize a 9VDC alkaline or nicad battery to supply enough current to maintain the memory until the power comes back on. Most backup batteries preserve the programming for several days or more. If the power outage lasts any longer, the chip goes blank and you must reset the time and reprogram the irrigation cycle(s).

When purchasing a digital controller, price should be the last consideration. Make sure you have the simplest-to-program model, not the cheapest. To compare prices, divide the number of stations into the total cost. Sometimes it's less expensive to buy, for example, two quality six-station timers than a single, more costly, 12-station model. In summary, before buying a digital irrigation clock, consider the following important points:
• Is the handbook readable, or does it require a Ph.D. in hieroglyphics?
• Are there enough stations to satisfy current and future irrigation needs?
• Can the clock be programmed for the longest anticipated irrigation run time?
• Can it be set to run for as short a time as you'll need for containers?
• How many stations to each program?
• Is the programming on an odd- or even-day basis, or both?
• Finally, what is the cost per station?

Multistation, Truly "Digital" Controllers

I've always considered myself to be a visual person. But I'm not a digital-readout kind of guy. The clocks I prefer have hands. Even though I have adjusted to the right-angled numbers in the digital LCD displays found in so many modern appliances, I'd much rather see hour and minute hands. Fortunately, for LCD-impaired people like me there are two styles of truly digital controllers—meaning that you use your little digits to set the timer. Although I have achieved a measure of victory over the all-electronic, digital LCD controllers, I've basically thrown in the towel and now demand an automatic controller which I can really get my hands on.

There is a style of old-fashioned controller which

has one or more rotating circles with various pins. The circles, or dials, have the time and day of the week printed around their circumference. These controllers, often called electromechanical controllers, are distinguished by their sturdy metal boxes and relatively straightforward instructions, except for discussions of "normally open" and "normally closed" circuits. This is the kind of timer my grandparents used to use, long ago, to control lights while they were on vacation. More recent models look similar but have—like the rest of contemporary merchandising—more plastic parts. To set such a controller, one simply pulls or inserts a pin, also called a tripper, for the start time and finish time. One nice feature is the fact that the program on the dial is not lost if the power goes off, since it's set with mechanical pins, not programmed onto a microchip. There are four major drawbacks to many of these old-fashioned, original, truly digital controllers. First, they can control only one or two stations per dial. You can set one dial to come on several or more times per day, but the start and stop times are wired to only one or two posts per controller. Second, because the old-fashioned electromechanical controllers worked on 110VAC and were designed to control 110VAC devices, you had to step the power down to 24VDC after the clock—an extra cost and hassle. Third, the minimum run time, due to the mechanical nature of the trippers, is usually quite long. One model I use with automated gray water systems runs the valve for two hours for each tripper inserted in the dial. Although this is fine with a gray water system, in which I want each valve to be open for 24 hours at a time, it will not work with most daily irrigation schedules. Finally, these controllers always cost more per station than their all-electronic cousins. The modern plastic- and metal-bodied versions have eliminated all the above-mentioned concerns and have

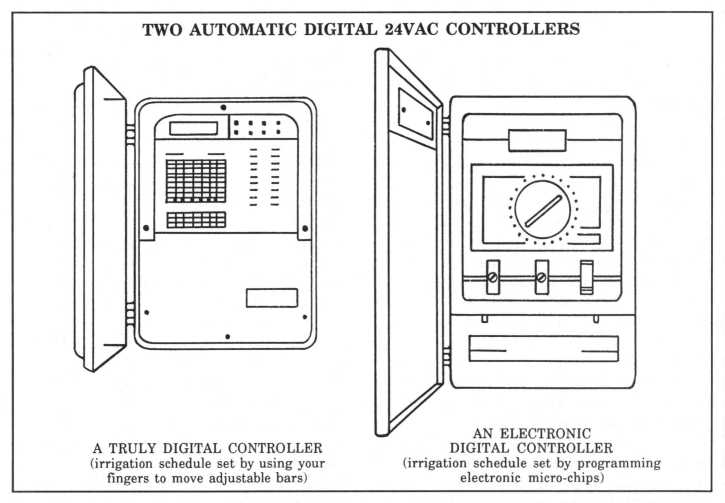

TWO AUTOMATIC DIGITAL 24VAC CONTROLLERS

A TRULY DIGITAL CONTROLLER
(irrigation schedule set by using your
fingers to move adjustable bars)

AN ELECTRONIC
DIGITAL CONTROLLER
(irrigation schedule set by programming
electronic micro-chips)

Figure 70 Two versions of automatic, multi-station irrigation controllers. The electronic version requires the gardener to be well-versed in the language of LCD readout. The other controller is also a digital device, but utilizes the gardener's actual digits for programming—the simple sliding of a bar sets the irrigation time for each station.

run times as short as one to three minutes. Most importantly, the modern electromechanical controllers step the power down to 24VAC for full compatibility with the solenoids for sale.

Fortunately, the irrigation industry has slowly begun to hear the plaintive cries of the LCD-impaired. A recent innovation has been a modern electronic controller designed to actuate 24VAC valves but with sliding bars for each station. **(See Figure 70.)** The sliding bars are like those adjustable levers on fancy modern stereos, amplifiers and radios by which the listener can fine-tune different frequencies of sound—often called graphic equalizers, even though no drawings are involved. The lever for each separate station is merely moved to the time indicated on the vertical axis. The irrigation settings on some of these clocks range from minutes to hours. This is my favorite style of controller because it's easy to see at a glance the setting for each station. Also, the length of irrigation can be reset almost effortlessly as frequently as the weather mandates. Although these units usually cost slightly more than the least expensive LCD controllers, the extra cost is quickly repaid in reduced aspirin consumption.

Super Controllers, for the Wealthy and Daring

For the ambitious and financially endowed gardener, there are plenty more options for greater complexity, enhanced features and a bigger dent in the budget. The following possibilities will remain dreams for most of us, things to consider in the future after our mastery of dripology has matured and our lottery ticket hits the big one. Some features available now and in the near future include the following:

• Periodic summer rains may temporarily negate the need for irrigation. Unfortunately, the clock on an automatic timer continues to tick. Normally, you'd have to remember to run out to the timer and throw the main switch from automatic to rain or off. With a rain sensor wired to the timer, this troublesome task becomes automated. Since I live where the seasonal drought is continuous, I've never tried rain sensors. If you'll be needing one, be sure the controller you purchase has this capability. Much can go wrong with these widgets, so don't buy one without a referral. Ask friends,

landscapers and suppliers for their personal experience with specific models.

• Soil water sensors, such as porous blocks or tensiometers, can be buried at one or more depths in the soil to monitor the irrigation water's penetration and the soil's degree of moisture. A simple hand-held gauge, costing $200 to $250, is connected to the porous block's ($5 to $6 each) wire leads where they come out of the ground, and the degree of soil moisture can be measured. Or, in the case of tensiometers, a probe is buried to a specific level and you visually read a calibrated meter. Tensiometers cost between $35 and $70 for the standard agricultural models. These same soil water sensors can be used to prevent the automatic timer from irrigating an already wet soil. The sensor is wired to the timer so that the current can't pass to its corresponding solenoid if the soil is wetter than a predetermined level. When the soil dries more than the preset level, the automatic station current goes directly from the controller to the solenoid. To explore these options, contact Irrometer Company, P.O. Box 2424, Riverside, CA 92516; Soilmoisture Equipment Corporation, P.O. Box 30025, Santa Barbara, CA 93105 or Rosemount Analytical Inc., 89 Commerce Road, Cedar Grove, NJ 07009.

• Large estates may benefit from having the controller located outdoors, in the landscape itself. This provides straightforward access for reprogramming and direct observation when running each station using the manual mode. These weatherproof, metal-bodied housings always add to the cost of the controller—from as little as $15 extra for painted metal boxes to as much as $2000 for stainless steel pedestal models, which look like a preacher's pulpit.

• If the hassle of manually turning on a system and running across your estate to see the station come on and watch for leaks is too much, technology has the solution. One manufacturer sells an adapter for certain controllers which allows the use of a hand-held remote control module. Much as with the remote controller for a VCR or television, you can stand just about anywhere in the landscape and turn on any desired station. This time-saving feature will set you back well over $2000. Not for everybody. (Mostly for the rich or the professional landscape service.)

• Since ET-based irrigation is the most accurate and efficient way to irrigate, why not adjust the irrigation time each day, as per the actual ET? You'll need a weather station, a PC computer and some software. Since most ET data are based upon a few fairly scattered weather stations, chances are there isn't a facility near you. (Each state is different. In California, more than 60 weather stations provide computer modem transfer of data on a daily basis.) For the most accurate weather information for ET calculations, the fanatic gardener would need a personal weather station capable of recording temperature, wind speed, humidity and rainfall. Automatic weather stations with paper or digital recorders start at $350 and go as high as $1400. But to utilize this raw information, you'll need a computer and software to crunch the numbers into an ET-based recommendation. Unfortunately, the only programs available are geared toward agricultural applications—which means they're far more powerful and costly than is appropriate for the residential landscape. Prices for PC-based software start at $500.

• In the not-too-distant future, we'll see personal computer controlled drip irrigation systems for the home in which weather data are automatically converted to adjust the duration of irrigation up or down. Right now, such a robotic system is designed only for the landscape and golf course industries and comes with a price tag of $5000 or more. More than the Cadillac of drip systems—the Lamborghini.

Until George Lucas has his people at Industrial Light and Magic turn to irrigation systems programmed by three-dimensional, virtual reality animated software which allows you to merely voice your commands, we'll just have to make do with the above wish list. In the interim, there are plenty of gizmos and widgets to confound and satisfy every gardener.

As the future comes rushing into the present, water is destined to be treated more and more as a nonrenewable resource, analogous to oil. With supplies dwindling relative to the population, water will become a political asset with layers of moral and ethical concerns. One recent film about two peasant families, two farms, a cistern and a spring has already portrayed our increasingly arid future. The French film *Jean de Florette*, by director Claude Berri, is set in the arid hills of southern France. In the story, a severe drought has dried out the neighboring farmer's cistern, and his crops and farm begin to fail. Two peasant farmers, played by Gerard de Pardieu (who later gained fame in *Green Card*) and Yves Montand, plot to force their neighbors into bankruptcy so their farm can be purchased at a great discount. The two farmers withhold their knowledge of a forgotten spring on the neighbors' property, using water as the ultimate silent weapon. Pardieu's and Montand's characters watch as the family struggles against drought, crop failure, meat rabbits dying from thirst and financial ruin. Alas, the two farmers manage to aid in the untimely death of the neighbor, to figuratively destroy his family and to acquire the property. The use of drip irrigation alone may not have been able to save this peasant family from devastation, but it is one of countless devices and strategies which can help prevent water shortages, strip the political overtones from water and preclude water from being used as moral or corporeal munition.

Appendix

APPENDIX #1: MAXIMUM LENGTH OF TUBING

This is perhaps the single most important illustration in the book: a listing of various drip irrigation tubing and pipes, their corresponding maximum effective lengths for even distribution of water and the corresponding maximum flow rate in gph. (The PSI column gives suggested maximum psi for each material.) Each listing has a "fudge factor" built into the recommended maximum length to compensate for pressure loss due to tees, elbows and other fittings. Instead of using complicated formulas for pressure loss, the home owner and landscape designer can simply look up the tubing or pipe and see the total length each valve or sub-assembly should have. These figures aren't fixed in concrete. For example, 1/4" Laser tubing is rated at a maximum length of 30 feet. If your paper plan for a flower bed shows a total length of 35 feet, you'll still probably get a fairly even distribution along the entire length. Please note that these figures are for a level garden; any significant slope will decrease the figures. If there is any slope, it would be better, in this example, to break the 35-foot length of Laser tubing into two separate sub-assemblies, each attached to an adequate supply line (as described below). But don't expect good results if you try to get away with both the maximum length, or a bit more, and the maximum flow rate.

This chart can also be used to double-check the size of a supply line to make sure it will provide enough water to each sub-assembly. For example, 3/4-inch sch 40 PVC pipe is rated at a maximum flow rate of 480 gph at a pressure of 40 psi. If your plan calls for the use of 1/2-gph in-line emitter tubing with 12-inch intervals, the maximum lateral or sub-assembly length is 326 feet, with a flow rate of 202 gph (326 feet X .62 gph). So a single 3/4-inch PVC pipe can supply two sub-assemblies of a maximum 326 feet each, but not three entire assemblies. This 3/4-inch pipe could supply two sub-assemblies and an additional sub-assembly or lateral(s) totaling no more than 76 feet. As another example, a length of 3/8-inch solid drip hose can supply up to 100 gph, or 183 feet of 1/4-inch Laser tubing (110 divided by .6 gph/ft.), providing no single Laser tubing lateral exceeds 30 feet.

Basically, this chart gives you the maximum length of tubing or pipe for a number of common materials, and your task is to divide and conquer. Sketch your entire irrigation plan, double-check the flow rates for each sub-assembly or lateral(s), divide the laterals or sub-assemblies into more sub-assemblies if the recommended maximum length was exceeded, then double-check the supply pipe(s) to make sure they can pass enough water to satisfy all the laterals or sub-assemblies if turned on at the same time. Finally, increase the size of the supply pipes if required or add another supply pipe to a portion of the garden.

MAXIMUM LENGTH OF TUBING
— ON FLAT GROUND —

DO NOT DESIGN SYSTEM FOR BOTH MAXIMUM FLOW RATE AND MAXIMUM LENGTH

TUBING / PIPE	size	PSI	MAX. GPH	Maximum length on flat ground
SPAGHETTI	1/8"	20	15	(5')
SPAGHETTI	1/4"	20	30	(15')
LASER TUBE™	1/4" 6" centers	10	1.2/ft.	(15')
LASER TUBE™	1/4" 12" centers	10	.6/ft.	(30')
LASER TUBE™	3/8" 12" centers	10	.6/ft.	(100')
POROUS PIPE	1/2" O.D.* 3/8" I.D.*	6-30	60—100ft.	(100')
POROUS PIPE	5/8" O.D.* 1/2" I.D.*	10	.8/ft.	(200')
SOLID DRIP HOSE	3/8" (400)	25	110	(120')
SOLID DRIP HOSE	1/2" (700)	25	235	(240')
SOLID DRIP HOSE	3/4" (900)	25	475	(300')
SOLID SCH 40 PVC PIPE	1/2"	40	250	(200')
SOLID SCH 40 PVC PIPE	3/4"	40	480	(300')
SOLID SCH 40 PVC PIPE	1"	40	780	(300')
SOLID SCH 40 PVC PIPE	1 1/4"	40	1200	(300')
TECHLINE™ IN-LINE EMITTER TUBING	1/2 gph 12"centers	25	.62/ft.	(326')
TECHLINE™ IN-LINE EMITTER TUBING	1 gph 12" centers	25	.92/ft.	(248')
TECHLINE™ IN-LINE EMITTER TUBING	1/2 gph 24" centers	25	.37/ft.	(584')
TECHLINE™ IN-LINE EMITTER TUBING	1 gph 24" centers	25	.55/ft.	(444')

Length scale (ft.): 10 20 30 40 50 100 200 300 400 500 600

*O.D.–outside diameter   *I.D.–inside diameter

APPENDIX #2: Drip Irrigation Hardware Flow Rates

(DOUBLE-CHECK WITH YOUR SUPPLIER. FOR ALL BALL-VALVES, GARDEN VALVES, FAUCETS, SOLENOIDS AND CHECK-VALVES, USE THE SAME FLOW RATES AS SCH 40 PVC PIPE)

PART	operating PSI	GPH (gpm)
PLASTIC IN-LINE ATMOSPHERIC VACUUM BREAKER	40	(7 gpm)
SENNINGER™ PRESSURE REGULATOR — LOW	25	(.1-8 gpm)
SENNINGER™ PRESSURE REGULATOR — MED.	25	(2-20 gpm)
ADJUSTABLE PRESSURE REGULATOR PENN 700™ 3/4"	35 (MAX.)	(4 gpm)
Y-FILTERS *		
AGRICULTURAL PRODUCTS SPIN CLEAN™ (150 mesh screen) 3/4"	40	(11 gpm)
AGRICULTURAL PRODUCTS SPIN CLEAN™ (150 mesh screen) 1"	40	(28 gpm)
ARKAL™ (140 mesh screen) 3/4"	40	(8 gpm)
ARKAL™ (140 mesh screen) 1"	40	(26 gpm)
AMIAD™ (155 mesh screen) 3/4"	40	(13 gpm)
AMIAD™ (155 mesh screen) 1"	40	(30 gpm)

*DO NOT USE MAXIMUM FLOW RATE WITH WELL WATER OR DIRTY WATER.

All plumbing parts have an upper limit to the amount of water which will pass through them. In a drip irrigation system, either the backflow preventer or the filter will be the bottleneck, the limiting factor, in the amount of water the system can distribute. This chart covers the range of flow rates for various name-brand drip irrigation parts. All flow rates are based upon the recommended and typical operating pressure as listed in the PSI column. The vertical lines represent the rate in gallons per hour, and the parenthetical figures are the rates in gallons per minute. With the Senninger pressure regulators, the gray bars don't start at zero because these regulators don't function unless there is a minimum flow. With these regulators, make sure the total discharge rate of the emitters equals or exceeds this minimum rate--for the low-flow version, it is 6 gph; with the medium-flow model, it is 120 gpm. For any device not listed on this chart, check with your supplier for the maximum flow rate.

APPENDIX #3: SPOT SPITTER FLOW RATES

Spot Spitter™ Pattern		PRODUCT COLOR CODE	FLOW RATE IN GPH @ 10 PSI			
			MINI	LOW	MED.	HIGH
360°		TAN				19.2 (24)
		BLUE			9.06 (13.38)	
160°		BLACK				12.4 (17.4)
		GREEN			9.3 (13.7)	
		AVOCADO		5.2 (8.1)		
		TERRA-COTTA	2.94 (4.32)			
90°		BROWN		5.5 (8.4)		
		GRAY	2.64 (3.86)			

Spot Spitters are especially useful when irrigating container plants or large hanging plants. Compared with drip irrigation emitters, these are not low-flow devices. But Spot Spitters are more efficient than most sprinklers and some micro-spray heads. The flow rate numbers in the chart are for a system with a 10 psi regulator. The parenthetical figures list the flow rates in gph at 20 psi.

APPENDIX #4: SOIL PERCOLATION RATES

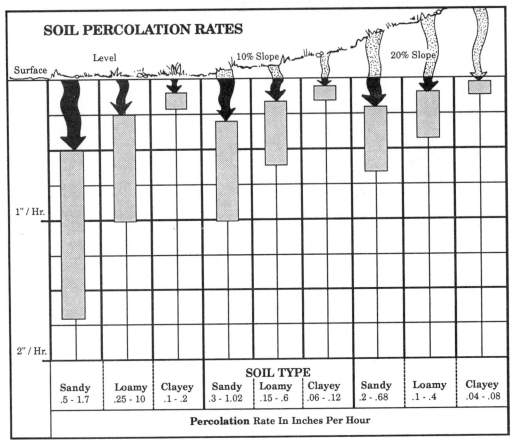

This chart reveals many things about how various soils respond to irrigation or rainfall. First, the graph represents the range of percolation in inches per hour. For example, a sandy soil will absorb from .5 to 1.7 inches of water per hour on flat ground. Second, the range of figures reveals that all sandy soils are not alike; there is a variation within each general category. Third, the chart shows how radically slope affects the percolation of water in different soil types. The same sandy soil which absorbed .5-1.7 inches on flat ground will absorb only .2 to .68 inch per hour on a 20% slope. The steeper the slope, the slower should be the flow rate of the emitters to prevent runoff.

APPENDIX #5: SOIL PENETRATION RATES

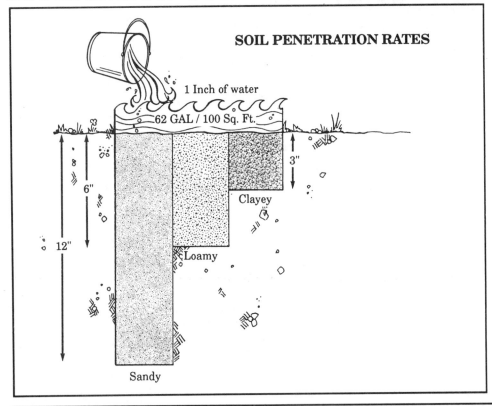

This is a graphic review of how different soil types absorb the same amount of water. Sixty-two gallons of water will cover 100 square feet to a depth of 1 inch. This amount of water will infiltrate only 3 inches deep in a clayey soil but up to four times deeper in a sandy soil. This chart shows why lower-flow emitters are important in clayey soils to prevent puddling, runoff and anaerobic conditions. This formula, 62 gallons per 100 square feet for each inch of water, can be used to convert irrigation recommendations based on inches into gallons for use with drip irrigation systems designed to moisten the garden's entire root zone.

APPENDIX #6: TRANSPIRATION WATER REQUIREMENTS OF ANNUAL CROPS

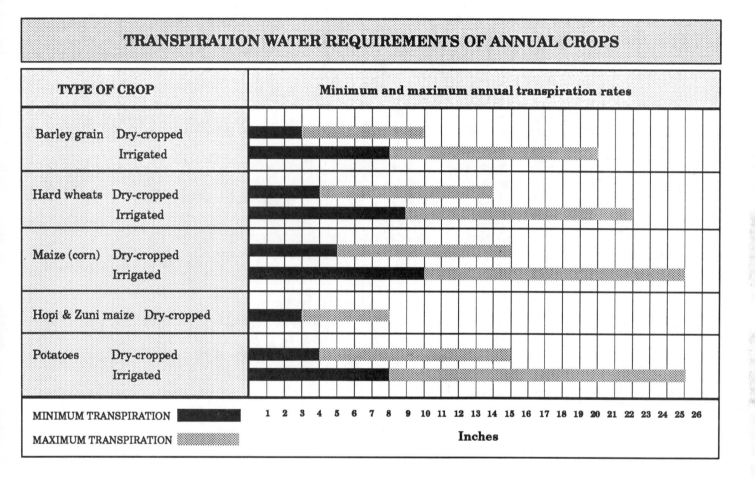

TYPE OF CROP		Minimum and maximum annual transpiration rates
TRANSPIRATION WATER REQUIREMENTS OF ANNUAL CROPS		
Barley grain	Dry-cropped	
	Irrigated	
Hard wheats	Dry-cropped	
	Irrigated	
Maize (corn)	Dry-cropped	
	Irrigated	
Hopi & Zuni maize	Dry-cropped	
Potatoes	Dry-cropped	
	Irrigated	
MINIMUM TRANSPIRATION		
MAXIMUM TRANSPIRATION		1 2 3 4 5 6 7 8 9 10 11 12 13 14 15 16 17 18 19 20 21 22 23 24 25 26 Inches

The black bar represents transpiration (inches per acre) used by the plant for growth up to the stage of producing a yield. The gray bar represents water used by the plant for producing the yield. Total water supplied to the crop must exceed the transpiration requirement by enough to cover losses from evaporation, runoff, and percolation below the root zone. Dry farmed crops transpire less because they are customarily planted farther apart, giving fewer plants per acre.

For detailed information on how crop yield is related to planting density and water consumption for these and many other crops, see "Water Requirements of Drought Resistant Plants," Johnson, 1977. Reprints are available from The Drip Irrigation Book Project, see Resources.

APPENDIX # 7: ANNUAL TRANSPIRATION REQUIREMENTS OF VARIOUS PERENNIAL CROPS

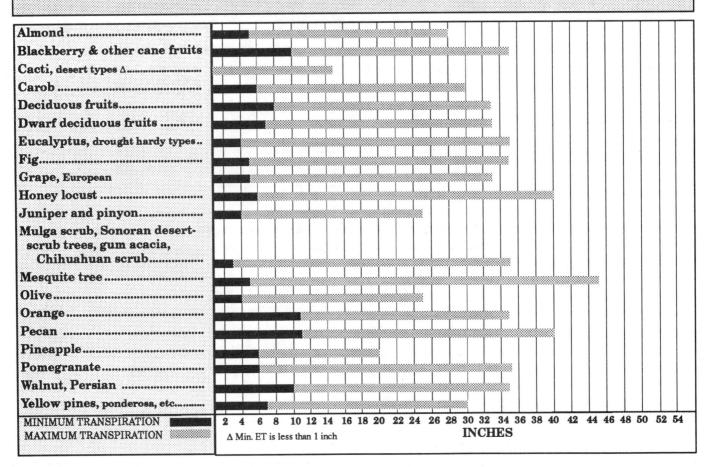

ANNUAL TRANSPIRATION REQUIREMENTS OF VARIOUS PERENNIAL CROPS

| Plant | MINIMUM AND MAXIMUM ANNUAL TRANSPIRATION RATES |

Plants listed:
- Almond
- Blackberry & other cane fruits
- Cacti, desert types △
- Carob
- Deciduous fruits
- Dwarf deciduous fruits
- Eucalyptus, drought hardy types
- Fig
- Grape, European
- Honey locust
- Juniper and pinyon
- Mulga scrub, Sonoran desert-scrub trees, gum acacia, Chihuahuan scrub
- Mesquite tree
- Olive
- Orange
- Pecan
- Pineapple
- Pomegranate
- Walnut, Persian
- Yellow pines, ponderosa, etc.

MINIMUM TRANSPIRATION ████
MAXIMUM TRANSPIRATION ░░░░

2 4 6 8 10 12 14 16 18 20 22 24 26 28 30 32 34 36 38 40 42 44 46 48 50 52 54

△ Min. ET is less than 1 inch   INCHES

The black bar represents the transpiration requirement (inches per acre) for survival, but little or no production, when the plants area planted far enough apart to eliminate root competition. The gray bar represents transpiration where the plants are planted closely enough that their canopies nearly touch, abundant water is available to the roots, and the climate is favorable. Total water supplied to the crop must exceed transpiration by enough to cover losses from evaporation, runoff, and percolation below the root zone.

For detailed information on how crop yield is related to planting density and water consumption for these and many other crops, see "Water Requirements of Drought Resistant Plants," Johnson 1977. Reprints are available from The Drip Irrigation Book Project, see Resources.

A Glossary of Water Words and Drip Stuff

actinomycetes — Soil inhabitants which resemble molds and fall between true molds and bacteria in classification. Their filaments, which are much smaller than a mushroom's, permeate the soil and form fruiting bodies. Actinomycetes thrive in moist, well-aerated soil and assist in the digestion of organic matter and lignin and the liberation of nutrients.

adapter — Generally speaking, any plumbing or drip irrigation part which connects one size of pipe or part to another. Often used to refer to the female fitting, whether glued or threaded, which joins different parts together. For example, a fht (female hose thread) X (by) fipt (female iron pipe thread) adapter converts hose threads to pipe threads. The male versions are often called transition nipples. (See transition nipple.)

aerobic soil — A well-drained soil with sufficient pore space to allow plenty of air to circulate in and out of the ground. The pore space is usually dependent upon a reasonable amount of organic matter and humus, both of which hold on to nutrients and keep the pore structure open. A healthy soil doesn't have to move nearly as fast as Jane Fonda to remain aerobic.

atmospheric vacuum breaker — This linguistic mouthful is actually a simple device which uses air to break any reverse siphon of water from the yard's irrigation system back into the home's water supply. Helps keep the home's water supply pure and disease free.

aviary wire — A wire mesh which resembles chicken wire in pattern but comes with 1/2-inch holes; the holes in chicken wire don't get smaller than 1 inch. Used to keep gophers out of raised wooden vegetable or flower beds and to line planting holes for shrubs and trees. Has a galvanized coating and may last up to ten years in the soil before rusting through.

backflow preventer — A device which keeps irrigation water from siphoning back into the house and contaminating the home's drinking water supply. There are three major types of backflow preventer: check-valves, atmospheric vacuum breakers and reduced pressure backflow preventers. Always check with your local building or water department to determine which backflow prevention device is code-approved for your area.

ball-valve — A valve which has a globe-shaped, rotating interior. The solid globe has a circular tunnel through one axis. When the handle of the valve is rotated, the solid portion of the ball cuts off the flow of water. Another rotation lines up the tunnel with the inside of the hose, and water flows through the valve. Ball-valves are often found at the discharge ports of quality Y-filters. Because ball-valves shear off any contaminants and because they don't easily wear out, as gate valves do, they are the preferred valve for manual irrigation systems.

barb — Not at all related to the famous stoic, anatomically incorrect child's doll. Any fitting with an arrow-like flange which allows the fitting to be inserted, like a fishhook in some poor catfish's throat, in only one direction. Barbed parts allow emitters to be inserted into the wall of solid drip hose and insert fittings to be placed inside the drip hose itself.

Bi-wall™ — Thin-walled plastic tubing used primarily in agriculture. Walls are too thin for typical landscape applications. Can't be curved — straight runs only. (See Mono-tubing and T-tape.)

check-valve — A backflow preventer with an internal spring around a plunger rod and a gasketed disk which stops any water siphoning back toward the house. Often not legal as the only backflow preventer in a drip system; must be coupled with some form of atmospheric vacuum breaker.

close nipple — A short, 3/4-inch pipe nipple (see nipple). Because the length is so short, close nipples are completely covered with male iron pipe threads. You'll never see one of these in the Sports Illustrated swimsuit issue!

combination stake — Used to hold up misters and spaghetti tubing in the landscape.

complex path emitter — See tortuous path emitter. (See, it wasn't that complex, was it?)

compression fitting — A type of fitting used to join two pieces of drip irrigation hose. The hose is forced inside a circular opening which has a barbed, ringed opening. Once inserted, the hose can't be easily pulled back out because of the circular barb. Unlike other types of fittings, this fitting seals better, up to a point, as the water pressure increases because the swelling hose squeezes tightly against the compression ring.

compression ring — A single compression barbed ring which can be glued into the slip (unthreaded) opening of a 1/2-inch PVC fitting to convert it to a drip irrigation compression fitting.

controller — An electronic device used to turn irrigation valves on and off automatically. Plugs into the home's 110VAC power and steps the power down to 24VAC to control the solenoid valves. Unlike timers, controllers usually service more than one irrigation line or zone. Most modern controllers are operated by computer chips and have digital readouts. As scientific as this sounds, controllers have a way of messin' up which remains unexplained, sporadic and very irritating. Besides, it takes a rocket scientist fluent in Gibberish to understand most of the manuals and to program the darn gizmo. (See also timer.)

coupler — Also called a coupling. Used to join two pieces of drip irrigation hose. Best done in the privacy of your own home.

crown — Does not appear when you spread margarine on the tree's bark. The crown of the root system is the upper 6 to 12 inches of the root system. This is the zone most vulnerable to crown rot (*Phytophthora* sp.).

crown rot — See *Phytophthora*.

D

diaphragm — Drip irrigation parts, in spite of their male and female names, don't need birth control. But some emitters use an internal diaphragm, usually made from some type of rubber, to modulate the flow of water and to help purge the emitter of any built-up sediment.

drain header — The drip hose, whether solid or with in-line emitters, which collects all the water at the end of the laterals farthest from the supply header. Like a supply header in that it's connected to one or more laterals. Usually a flushing end cap or a threaded end cap is added to the drain header to make it easy to flush the hose and drain out water for winterization.

drip — (1) An unwanted droplet of water at the base of your water heater which means a multihundred-dollar visit from the plumber. (2) The tiny dribble of water passing from an emitter. (3) That ugly oaf who sat behind you in ninth grade science class. (4) A style or technology of irrigation in which a tiny trickle of water is slowly applied to every plant.

drip hose adapter — The first fitting at the beginning of a drip irrigation system. Almost always an fht (female hose thread) swivel X (by) drip hose adapter. The female hose threads of the swivel go onto the male hose threads of a faucet or a transition nipple. The swivel action makes it easy to add or remove this fitting quickly. The other side of the adapter is either a slip (glue), compression, insert or Spin Loc™ part, depending upon the system.

drip line — A length of solid drip irrigation hose or in-line emitter tubing.

dripline — The width of a tree or shrub's foliage, where water would drip off the edge of the canopy. Not an indicator of the width of the root system—roots grow from one-half to three times wider than the dripline.

dripology — The modern, albeit inexact, science of drip irrigation. The only recognized school of higher learning for students of dripology is the School of Hard Knocks.

drought — Practically speaking, not a clearly defined word. The dictionary defines it as "a prolonged period of dry weather; lack of rain." But drought is certainly a relative term. A five-week dry spell in the humid South may be devastating to certain unirrigated plants, while five months of no rain whatsoever is routine in California.

dweeb — An accurate description of some plumbers— looks sort of like the school nerd and acts as if he or she has been around too much pipe dope; a certified klutz. (See klutz.)

E

elbow — As amazing as it may seem, this has nothing to do with that joint halfway between your wrist and shoulder. An elbow is actually a fitting which allows drip hose or pipe to make a 90° turn. Also informally called a "90." Nobody knows what the joint in the middle of your arm is really called.

elbow grease — My grandfather, "Popper," always told me, "A little elbow grease'll make the job go faster." I looked for years in his garage for a can or jar of this important compound, but to no avail. If you know where to get some, let me know.

emitter — (1) The little gizmos attached to, or built within, solid drip irrigation hose which control the flow of water to the landscape or plant(s). There are many name-brand versions of emitters, but they basically fall into four generic styles or technologies: single diaphragm, double diaphragm, tortuous (or complex) path and simple orifice. (2) A New Age self-help and personal growth addict who is always emitting his or her feelings.

end cap — The fitting added at the end of a drip hose lateral to make it easy to open the tubing for drainage or flushing. Has a female hose thread cap with a washer which threads onto the male hose thread fitting. The other end will be a compression, insert or Spin Loc™ opening, depending upon what system you have.

evapotranspiration — Loss of water from a plant or crop via transpiration from the foliage and evaporation from the soil's surface. The ET (evapotranspiration) rate is influenced by humidity, rainfall, wind speed, temperature and mulch.

F

faucet — The standard gizmo on the end of the pipe sticking out of the house's exterior wall or on top of a metal water pipe in the yard and onto which the garden hose is attached.

fertilizer injector — See proportioner.

fht — Plumbing moniker for a female hose thread.

figure-8 end closure — A simple end closure which involves threading the end of the drip hose through one side of the figure-8, bending the end of a drip hose over and securing the bent end inside the other half of the figure-8.

filter — A device with a metal screen (cheap, poor-quality models have plastic screens) which is used to trap any particulates, dirt or scum before it can clog the emitters. An essential component of all drip irrigation systems.

fipt — Plumbing moniker for a female iron pipe thread.

free lunch — No such thing.

G

gate-valve — A valve where the internal mechanism is a vertical-closing panel (gate). Because the gate rubs against its vertical channel, this type of valve is more likely to start leaking than a ball-valve. Usually has a round handle perpendicular to the valve's shaft.

gizmophobia — A recently defined modern illness. Symptoms include refusing to program the clock on your VCR; making coffee the old-fashioned way, with boiled water from the stove top; the reluctant ownership of an answering machine (severe gizmophobia involves refusal to purchase an answering machine and refusal to talk to such devices) and avoidance of all ATMs. Thought to be due to a virulent form of computer virus. No known cure.

goof plug — No dripologist should be without a set of goof plugs. This essential part is merely a solid barbed plug or barbed cap which is used to patch holes where emitters have been removed. The goof plug is inserted into the hole and left there permanently.

green manure — A cover crop that is tilled into the soil to improve its tilth and to act as a mild fertilizer. Has nothing to do with the color of any animal's feces.

hardware cloth — Not a cloth at all but a mesh of galvanized metal wire. Good for excluding gophers from plantings or lining the bottoms of irrigation main assembly boxes so the gopher or mole throws (piles of dug-up tunnel dirt) don't fill up the enclosures. Comes in a mesh with 1/8-inch to 1-inch or larger square holes.

hose-bib — See faucet.

hose shut-off valve — A small ball-valve which can be added at the end of a hose to control water without you having to run back and forth to the faucet. With a few extra parts, this valve can be spliced into any drip hose, allowing the gardener to exclude water from portions of a system. Often used with a drip irrigation system for vegetables.

in-line emitter hose — A more recent and effective type of drip irrigation hose in which the emitters are placed inside the hose at regular intervals by the manufacturer. The prespaced emitters use a tortuous path technology for water regulation without clogging. Water can be distributed at 1/2, 1 and 2 gph rates at many separate intervals ranging from 12 to 72 inches.

insert fitting — These fittings have male-shaped parts with barbed exteriors which insert inside the drip irrigation hose. As the water pressure increases, up to a point, the fitting is more likely to fail because the swelling drip hose can bloat away from the barbed posts. A ring clamp must be used to secure the hose against too much pressure.

IPS flex PVC — A thick-walled, flexible tubing which can be glued directly into 3/4-inch PVC slip fittings. The flexibility of the "pipe" makes it ideal for situations in which a riser coming out of the ground is likely to be knocked around or disturbed.

J-stake — A landscape pin used to secure drip irrigation hose, landscape netting and 12VDC wiring. Made like an upside-down version of the letter "J." Not as sturdy as the best U-stakes.

klutz — A cross between a twitching-schlemiel and a stumbling-politician. Someone who has plenty of coordination problems but no corporeal medical problem to blame them on. The modern word for a spaz.

Kourik — A bohemian name of obscure origin. The author is the last of the species. Not related to Katie Couric of television fame, yet.

labyrinth — A complex, tortuous path inside certain emitters. The labyrinth of passages keeps any sediment in the water in suspension so that it will pass through the emitter's orifice. All in-line emitter tubing uses some form of labyrinth to allow for a relatively large emitter orifice and to keep the emitter from clogging.

Laser tubing™ — Resembles 1/4-inch or 3/8-inch spaghetti tubing but has a small cut every 6 or 12 inches, depending upon the type. Much cheaper than in-line drip hose but will clog easily, is not pressure compensating and can't be used for very long runs.

Laser tubing™ adapter — Looks like a double goof plug, but there's a hole through the middle. Used like a transfer barb when attaching a lateral of Laser tubing™ to 1/2-inch drip irrigation hose. One end has a 3/8-inch barb, and the other end has a 1/4-inch barb. Can be used to adapt 1/4-inch spaghetti tubing to 3/8-inch Laser tubing™ or porous pipe.

lateral — A lateral is a water-bearing pipe or hose which originates as an offshoot of a main supply pipe. Laterals are usually attached to the supply line via a tee.

main assembly — The collection of parts at the main water supply (often a faucet) which keeps the home's drinking water pure, filters the water supply to the emitters and regulates the water pressure to keep the drip system intact. The main assembly is usually composed of a backflow preventer, filter and pressure regulator—plus the miscellaneous part needed to connect everything.

mesh — Most irrigation filters are rated by mesh size. The larger the mesh number, the better the filtration—smaller particles are trapped. Many metal screen filters are either 60 mesh (254 microns, or .010 inch), 100 mesh (152 microns, or .006 inch), 140 mesh (104 microns, or .004 inch) or 250 mesh (61 microns, or .0024 inch).

mht — Plumbing moniker for a male hose thread.

micron — A common measurement for irrigation parts. The bigger the micron number, the bigger the opening. A single micron equals one-millionth of a meter. Two hundred and fifty-four microns equals .010 inch and equals a 60 mesh screen.

mipt — Plumbing moniker for a male iron pipe thread.

Monotubing™ — The lowest cost thin-walled agricultural tubing; also the cheapest. Easily clogs and is often damaged in a landscape setting. Can't be adapted to curved plantings (See Bi-wall and T-tape).

nipple — Actually, not as sexy as it might sound. Plumbing nipples come in plastic and iron versions, with male iron pipe threads on each end. Unlike people's, the plumbing nipple ranges in size from 3/4 inch (see close nipple) to 48 inches in length. Used to join two female iron pipe threads.

Numero Uno — The first, most frequently occurring cosmic plumbing law. See PMLP and page 4.

O-ring — Not a zero ring but a circular rubber gasket. Used in a union fitting to seal the joint where the two flush plates of each side of the union meet. Not attached to the union—don't let the O-ring fall out or disappear.

Phytophthora — The Latin genus name for various species of fungal diseases which attack the upper portion of the roots to destroy the bark's active layers of transport. Often called crown rot.

pipe dope — The sticky, gooey stuff you spread on pipe threads to make sure there are no leaks. Pipe dope has a magical magnetic attraction to all clean clothes, rugs and expensive suits. Other than the mess, any dope can use the stuff.

PMLP — Pop Murphy's Laws of Plumbing. A list of the seven most annoying, yet universally common, plumbing problems, as inscribed on a pair of galvanized tablets. Developed and inflicted by Pop Murphy. These rare tablets were uncovered at a secret, undisclosed site at a large urban hardware store beneath an enormous pile of old, corroding and rusting special-order plumbing parts which were never delivered. (See also Pop Murphy.)

Pop Murphy — The inventor of devious, cunning and irritating plumbing gizmos. Also the perpetrator of all leaks, broken fittings and plumbing problems. Honored in the recently found tablets titled "Pop Murphy's Laws of Plumbing." (See PMLP.)

porous hose — Unlike an emitter, in which the water dribbles out at select points, porous drip hose is designed so that water oozes out through the entire surface area of the hose's walls. A rather old drip irrigation technology and one which is mostly used with residential, not agricultural, systems. This genre of drip hose works well only with chlorinated city water because it's so prone to getting clogged by sediment and becoming sealed off internally due to the buildup of various types of algae slimes.

pressure-compensating emitter — A special type of emitter which is engineered so that the flow rate stays the same regardless of the length of the line (up to a point) and any change in elevation. Required when irrigating landscapes with a total elevation change of 20 feet or more. Has nothing to do with stress reduction—you're on your own.

pressure gauge — A handy device which is attached to a faucet or to the end of a drip line to check on the operating psi (water pressure, in pounds per square inch).

pressure regulator — A gizmo which magically reduces the water pressure in the city or home's plumbing to 25 psi or lower to protect the subsequent drip irrigation fittings. Put one in with every main assembly. Don't ask how it works; just do it.

proportioner — A special device which draws a proportioned amount of a concentrated solution out of a tank and mixes it with the irrigation water going through the drip system. Can be used to apply soluble fertilizers or to clean out the tubing and emitters with dilutions of acids.

psi — Pounds per square inch; the unit of measure for water pressure. Typical home water pressure is 40 to 80 psi. Drip irrigation systems generally operate at 10 to 25 psi.

punch — A hand-held device used to poke a small hole in solid drip hose so an emitter can be added to the hose. Not required with in-line emitter tubing.

PVC — A type of semi-rigid plastic made from polyvinyl chloride (PVC) which is often used for garden plumbing. Some of the more common grades of this pipe are (from sturdiest to weakest walls) schedule 80, schedule 40, class 200 and class 120, which resist bursting, in 1/2-inch to 1-inch pipe, up to, respectively, 620-850, 370-480, 200 and 120 pounds per square inch.

reduced pressure backflow preventer — The best and most expensive type of backflow preventer. Best plumbed in at one location to protect the home's water supply from all of the exterior irrigation lines. Must be installed by a professional plumber or landscape contractor.

riser — The vertical pipe coming out of the ground to a faucet. Actually a very long nipple. (See also nipple.)

saline water — Irrigation or groundwater which is high in salt (sodium chloride). While saline water is useful in many medical applications, it is not healthy for many plants.

sch — Stands for the word "schedule." Used to denote the type or grade of PVC pipe and fittings—sch 40 or sch 80.

shrubbery (also shrubberry) — Plants which are smaller than most trees (and dumber) and bigger than ground covers (but slower, in every way) but refuse to get a real life. Also known as one of the prerequisite offerings for travelers who wish to appease the guardians of the Dark Forest of Evening — the Knights-Who-Say-Kneek — and pass through those enchanted woods unharmed. (See the movie *In Search of the Holy Grail*, by Monty Python.)

sinker roots — Those tree roots which grow straight down and help anchor the tree against wind, snow and ice. Not the same as taproots. Sinker roots descend into the ground from all areas of the root system.

slip — It would be nice if this word meant that your garden (or mine, for that matter) were littered with lots of skimpy underclothing. Alas, it just means a PVC fitting with an opening which requires glue, as opposed to threads with pipe dope, to "weld" the two parts together. Usually, the end of the hard, rigid PVC irrigation pipe and the fitting are moistened with PVC glue and the pipe is slipped into the wet, round opening of the waiting fitting.

solenoid — The electrical valves used to control drip irrigation systems. The wires to the electrical valve usually carry 24 volts of AC power. The irrigation controller has a transformer to step down the house current to 24VAC. These valves are dependent upon the static line pressure of the water supply to assist in the opening and closing of the valve.

spaghetti stake — Either a slotted stake that holds the spaghetti tubing in place in the landscape or a plastic stake with a post onto which the spaghetti is slipped, turning the stake into a form of emitter.

spaghetti tubing — Don't try to eat this stuff at home. Actually, a tiny or slender type of polyethylene tubing which can be used to distribute water to emitters or plants. Comes in 1/4- and 1/8-inch diameters. Because of this tubing's propensity to knot up, twist around itself and—in the words of Joe Bob Briggs—"make the sign of the double pretzel," it appears to be breeding and will most surely make a tangled mess of itself. Only a severely masochistic person would use this stuff out in the landscape. It can be controlled with its use in container plantings.

speed coupling — A fitting used with garden hoses and devices with hose threads that allows for the quick insertion (coupling) of two parts instead of the tedious process of screwing the parts together. The coupling mechanism resembles the male and female fittings found on an air compressor hose at the gas station. (Although, with the status of the modern American filling station, you're lucky to find an operating version.)

Spin Loc™ — A series of fittings that allow for easy connection of drip irrigation hose. Utilizes a male post that the drip hose fits over and a threaded ring that tightens down over the outside of the hose.

spot emitter — Not a true emitter because it makes a spray. A plastic stake used with spaghetti tubing. The spaghetti is inserted onto a post at the top of the stake; the depth and width of the V-cut in the post control both the flow rate and the spray pattern.

sprinkler — A soon-to-be-antique method of irrigation which pisses a large amount of the irrigation water into the wind.

stabilizing leg — For people who drink and irrigate. Not really. Metal legs attached to a main assembly so it isn't stressed as it hangs off the faucet and won't be broken if bumped into.

sub-system — A branched system of laterals originating from a main supply line. Unlike a single lateral, a sub-system, also called a sub-main, usually has several subordinate lines all connected by tees in a pattern similar to a candelabra.

supply header — The solid or in-line drip irrigation hose which supplies one or more laterals.

swivel — A rotating fitting that can be screwed onto another fitting. Usually refers to a female hose thread (fht) which is threaded onto the end of a hose, faucet or drip irrigation part. Usually requires a rubber gasket in the swivel to prevent leaks.

T-tape™ — Thin-walled agricultural product which clogs readily and is too easily damaged for landscapes (See Mono-tubing and Bi-wall).

tee — A fitting which joins a lateral line (solid PVC pipe, in-line emitter tubing or solid drip hose) to another water supply line. Tees come in compression, slip (glued), barbed-insert or Spin Loc™ models. In this book, a device which has nothing to do with getting a tiny ball to disappear into a hole in the ground.

timer — A battery-powered controller which controls one irrigation line. Attaches to the faucet and controls the flow of water to a hose or drip irrigation system.

tortuous path emitter — Emitter that contains a complex path or labyrinth which allows larger particles to flow through it without causing clogging. This type of emitter stays virtually clog-free, even when used with well water or filtered water that remains high in sediments and suspended solids or when iron oxides, dissolved calcium and other minerals are present in the water supply. One of the more recent developments in drip irrigation technology and one which has virtually eliminated the chance of clogging—the Achilles heel of the older types of emitters

transfer barb — A plastic part usually used to connect spaghetti tubing or Laser tubing™ to solid drip hose. The drip hose is first pierced with a punch to make a hole. Then, one of the rigid posts with a barb is inserted to keep the hole from squeezing shut around the spaghetti or Laser tubing™. The other barbed post secures the spaghetti or Laser tubing™. (See also Laser™ tubing adapter.)

transition nipple — A plastic or metal fitting with a male hose thread (mht) and a male iron pipe thread (mipt) which is often used to connect conventional garden plumbing to drip irrigation fittings.

U-clamp — You clamp, I clamp—what the heck, we all clamp. The metal strap which is used to mount pipes to walls, decks, joists or posts. Doesn't really look like a U, but who's asking?

union — Actually, somewhat related to a coupling. A union is a plumbing part which, after the locking ring is unthreaded, separates into two pieces and allows you to take a portion of any irrigation system (providing there is an union on each end of the section) out for repairs without having to cut the pipe. The use of unions allows for the quick reinstallation of the repaired section without your having to reglue with extra fittings.

U-stake — A landscape pin used to secure drip irrigation hose, landscape netting and 12VDC wiring. Shaped like an inverted letter "U." Much sturdier than a J-stake.

variable spaghetti stake — A plastic landscape stake with a V-shaped opening to adapt to various sizes of spaghetti tubing or Laser tubing™. Ideally, it attempts to hold the tubing in place, but it is often unsuccessful in a real landscape. Best used with container plants.

W

wet spot — Has nothing to do with potty training or sexual politics; the often ill defined extent of moist soil beneath each emitter. The wet spot in drip irrigation, unlike other wet spots, has both depth and breadth—the extent of which is dependent upon the rate of the dribble

(in gph), the duration of the trickle (in hours), the soil type, the slope of the land and the climate.

who-ha — (1) An unlabeled plumbing fitting of unknown purpose. (2) One-half as creative as a nerd but smarter than a dweeb. (See dweeb.)

X — In spite of the lascivious nature of male and female plumbing fittings, there are no X-rated plumbing parts. Actually, a plumbing symbol for the word "by," used to denote a fitting's specifications. The notation "mipt X mht" reads "male iron pipe thread by male hose thread"—which is a transition nipple. (See also transition nipple.)

Y-filter — Because. Actually, the best type of filter for a drip irrigation system. Easily identified by the filter chamber, which is integrated into the filter at an obtuse angle. The best Y-filters have a metal screen filter within the filter chamber and a ball-valve at the end of the chamber to make it easy to flush the screen.

Y-valve — A valve which has two hose hookups. It allows two hoses or irrigation devices to be attached to one faucet. Usually has a small ball-valve on each of the two male hose thread (mht) hookups.

Z

z end.

Suppliers and Resources

Suppliers

I always recommend finding a local drip irrigation hardware supplier over a mail-order company. Find a local drip irrigation company with whom you can ask any and all questions. Unfortunately, quality drip irrigation hardware, especially in-line emitter technology, is hard to find in many towns and cities around the country. Therefore this listing of mail-order sources for drip irrigation widgets.

Most gardening mail-order companies sell a variety of irrigation products. A majority of these catalogs do not offer in-line emitter tubing and the requisite fittings. Since I'm biased toward in-line emitter technology, I've chosen not to list all mail-order sources of every manner of irrigation hardware. Instead, what follows is a basic listing of those companies which offer irrigation gizmos including in-line emitter tubing. This listing is not meant to be exhaustive, but even the most ambitious dripologist will be able to satisfy his or her drip techno-fix.

The Gardener's Supply Company
128 Intervale Road
Burlington, VT 05401
(802) 863-1700
(802) 660-4630 *(Irrigation Catalog)*

Has an extensive gardening mail-order catalog with quality no-nonsense, get-down-and-get-dirty gardening tools. Has recently started to offer some in-line emitter tubing, but it's not listed in the large, full-color catalog, which features porous pipe. Call or write and ask for the *Irrigation Catalog* with a listing of in-line emitter tubing. Both catalogs are free.

Harmony Farm Supply
P.O. Box 460
Graton, CA 95444
(707) 823-9125

Has a large inventory of many types of drip irrigation and sprinkler parts and stocks several configurations of in-line emitter tubing. The comprehensive catalog is best suited to those with some previous irrigation experience, but everyone will benefit from reading this catalog. Also includes organic gardening and farming supplies, seed, books and tools. Has a large, spacious retail store located in Sonoma County wine country. Catalog is $2.

The Natural Gardening Company
217 San Anselmo Avenue
San Anselmo, CA 94960
(415) 456-5060

Hey, what can I say—a great catalog. The only source of the official Robert Kourik Drip Irrigation System. The system comes as a kit composed of all the best parts I've found over the past 12 years. Each part in the main assembly comes from a different manufacturer and was selected to be the sturdiest available. The main assembly comes preassembled with Teflon tape, so all you have to do is thread it onto a faucet. You can also purchase each part or fitting by the piece. They also offer a complete array of quality gardening tools, gourmet vegetable seed and wildflower seed, organic gardening fertilizers and pest controls and the country's only mail-order source of organically grown herbs, flowers and vegetables. The mail-order and retail store are located at the same address. Catalog is free.

The Urban Farmer Store
2833 Vicente Street
San Francisco, CA 94116
(415) 661-2204
(800) 753-3747 (Order line)

A comprehensive offering of automatic and manual drip irrigation systems. Carries in-line emitter tubing from several manufacturers and all the required fittings. Ask for the special sheet listing the in-line emitter tubing. Oriented to the urban and suburban gardener, not the commercial farmer. Has a 6000-square-foot retail store near Golden Gate Park. Helps sponsor the prestigious annual San Francisco Landscape Garden Show. Catalog is $1.

Resources

These are some of my favorite books, magazines and catalogs related to irrigation and water in general:

"Water Requirements of Drought-Resistant Plants"
by David E. Johnson
Occidental, CA: Metamorphic Press, 1977 (reprinted 1992)

The transpiration charts for annual crops, trees and shrubs in the Appendix were based on this fascinating booklet. The complete paper includes a thorough discussion of transpiration, ET rates, crown transpiration and root transpiration. The focus is on dozens of food and economic crops for arid lands. The intent of the paper is to provide farmers with guidelines for the economic potential of various crops under dry-farmed and irrigated conditions, to show how irrigation affects the yields per acre. This ground-breaking literature research has not, to my knowledge, been superseded by any other ET literature since its original publication date. Because I found this paper and its six charts and graphs so intriguing, I contacted Dave and asked to reprint it for readers of this book. He agreed. A photocopy of this 21-page paper is $6 (postage, handling and CA sales tax included). Write the check to The Drip Irrigation Book Project and send to P.O. Box 1841, Santa Rosa, CA 95402.

Sensitive Chaos
by Theodor Schwenk
New York: Schocken Books, 1976

If you want to get in touch with the essence of water, this is the book. The text and many drawings and photographs show how water's movement reveals a universal design. The patterns of movement for both minute and macro streams of water are clearly illuminated. Explains how energy is stored and dissipated as water attempts to flow toward an universal equilibrium. This information explains the flow of silt in a gutter, mud in a river or sediment within an emitter. The presentation of water's (and smoke's) propensity to form a vortex is both fascinating and poetic and reflects what happens in the complex tortuous path of an in-line emitter. The correlation between water's primordial movements and the early formation of an embryo is particularly engaging.

"Patterns of Swirls in Air and Water"
by Hans J. Lugt
Orion, *Nature Quarterly*, Winter 1985

A more photographically lyrical version of the same concepts, patterns and dynamics discussed in *Sensitive Chaos*. Even links the swirling vortex of water and wind to certain Vincent Van Gogh paintings and *Voyager* photographs of the Great Red Spot on Jupiter. A preceding article, "A Magic Ratio Recurs Throughout Art and Nature," discusses the mathematics of spirals and vortexes and the Fibonacci sequence, or golden (logarithmic) spiral, and how they are manifest in nature—from shells to sunflower blossoms to pine cones.

Cadillac Desert
by Marc Reisner
New York: Viking Penguin, 1986

Still the definitive lay book on the complex politics and sociology of water in the arid West. One of the first books to reveal the extent of groundwater overdrafting (excessive pumping), land subsidies, state and federal aqueducts and the implications for our future. One of the first books to reveal how water in the West is more a nonrenewable resource than a limitless given.

Overtapped Oasis
by Marc Reisner and Sarah Bates
Washington DC: Island Press, 1990

The sequel to *Cadillac Desert*. Focuses on the politics of water systems and goes into great detail about the patchwork of laws which govern water's use throughout the West. Basically an extensive position paper on the use of water marketing (sales of water without loss of water rights) as a solution to the conflict between agricultural and urban users. Presents a chapter of detailed recommendations for modernizing the archaic hodgepodge of western water sources, laws and distribution channels. More technical than Reisner's first book, not casual reading—but an eye-opener.

Killing the Hidden Waters
by Charles Bowden
Austin: University of Texas Press, 1977

A tragic and revealing portrait of the Papago Indians, who inhabited the Sonoran desert and lived a rich life in an environment which receives only 3 to 10 inches of rain per year. These nomadic people started keeping their history on carved and notched sticks in the early 1800s. The sticks fail to mention the Mexican War or the gold rush of 1849, but they mention various thunderstorms and—the beginning of the end for them—the drilling of the first well by white men in 1912. This book reveals the pathos and bitter irony of a native people who, after generations of living within the natural limits of their ecosystem, watch and participate in the destruction of their own culture with the introduction of wells and pumps. The turning point comes when the tribe's chief is found, late at night, drinking from the very well he has forbidden others to use. An excellent metaphor for our modern culture in the arid West.

The Water Books Catalog
AgAccess
P.O. Box 2008, Davis, CA 95617 (Mail order)
603 Fourth Street, Davis, CA 95616 (Retail store)
(916) 756-7177

At last, a catalog devoted exclusively to books about water. All topics are covered, including water purification, irrigation, policy and law and water conservation.

The ABC's of Lawn Sprinkler Systems
by A. C. Sarsfield
Irrigation Technical Services, P.O. Box 268, Lafayette, CA 94549

Thoroughly covers fixed-pipe sprinkler systems for lawns. Explains both fixed-head and pop-up sprinklers,

all design procedures, step-by-step installation techniques, how to calculate pressure losses and tips for automating the system. The book is rather detailed and technical, is somewhat dry and well illustrated and has detailed pressure loss charts in the appendix. May put off beginners at first, but it's the best book on the topic.

Root Development of Field Crops
by John Weaver
New York: McGraw-Hill, 1926

Root Development of Vegetable Crops
by John Weaver
New York: McGraw-Hill, 1927

These out-of-print books are the best sources for graphs of root development based on actual root excavations. Amazing patience and attention to detail went into the research behind these books. They dispel many myths about where to water and fertilize plants by graphically showing the real root zone. The basis for many of the root zone drawings in my book on edible landscaping. Charts show lettuce roots to 7 feet deep, spinach roots to 3.5 feet, the fibrous roots of a cauliflower and vining winter squash with 25-foot-wide root systems. (A reprint of the best root zone charts from these two books and other papers and articles in my files is being considered. To be notified of publication of the root zone handbook, send a postcard requesting notification to Root Zones, P.O. Box 1841, Santa Rosa, CA 95402.)

"Cisterns Deliver the Rain Water"
by Robert Kourik
Garbage magazine, July/August 1992

Cisterns can capture free rainfall and provide water for drip irrigation. A 1000-square-foot roof gathers 620 gallons of water with each inch of rain. Cisterns are a relatively simple technology which can give a home owner a measure of water independence—and plants prefer rainwater over chlorinated city water. (This article is a brief overview of my forthcoming handbook on the subject. To be notified of its publication, drop a postcard to Cisterns, P.O. Box 1841, Santa Rosa, CA 95402.)

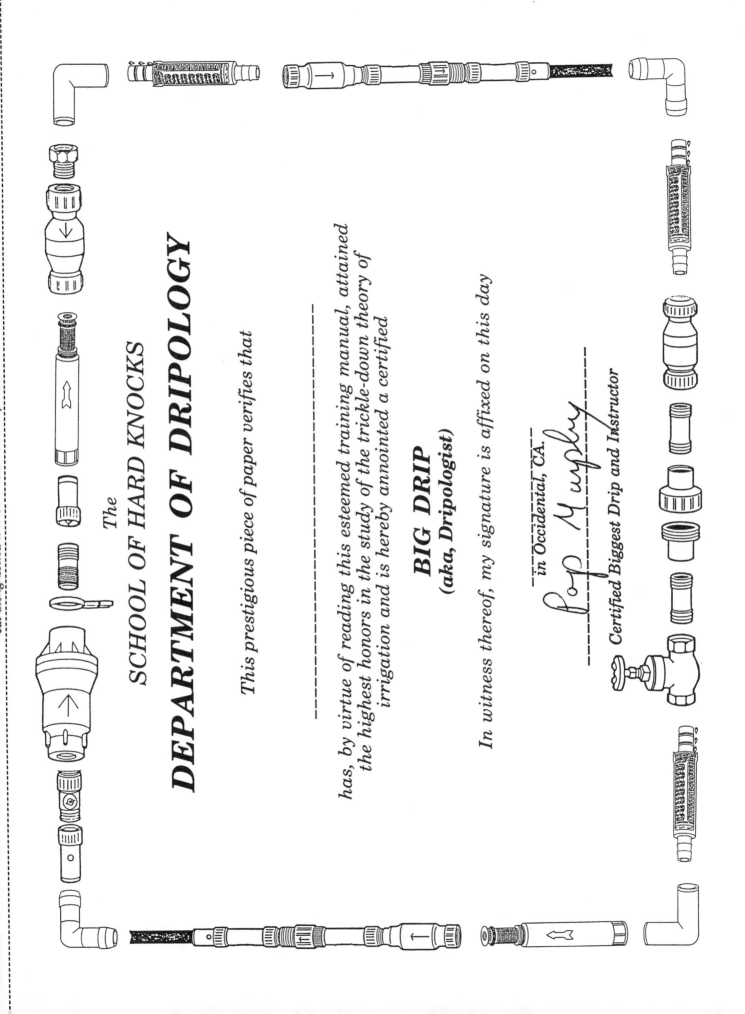

The

SCHOOL OF HARD KNOCKS

DEPARTMENT OF DRIPOLOGY

This prestigious piece of paper verifies that

_____

has, by virtue of reading this esteemed training manual, attained the highest honors in the study of the trickle-down theory of irrigation and is hereby annointed a certified

BIG DRIP
(aka, Dripologist)

In witness thereof, my signature is affixed on this day

_____
in Occidental, CA.

Certified Biggest Drip and Instructor

Designing and
Maintaining

YOUR EDIBLE LANDSCAPE NATURALLY

by Robert Kourik

Now, after more than six years in print,
A CLASSIC REFERENCE BOOK
FEATURING:

384 PAGES, 8 PAGES FULL COLOR PHOTOGRAPHS, 103 LINE DRAWINGS, 65 CHARTS

from the foreword by:

ROSALIND CREASY, author of <u>The Complete Book of Edible Landscaping</u> and <u>Earthly Delights</u>

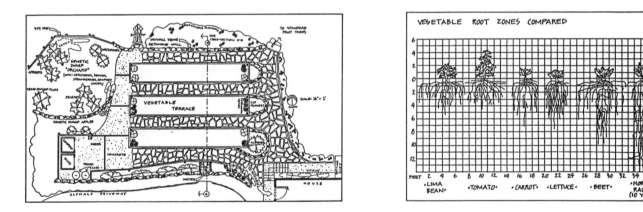

Robert Smaus, Horticultural Editor for the Los Angeles Times,
has rated *Designing and Maintaining Your Edible Landscape - Naturally*
as

"One of the nine most essential books for west coast gardeners."

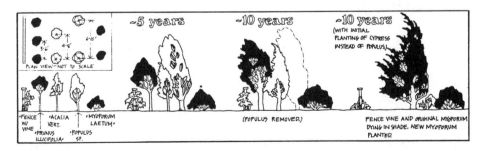

Gardeners throughout the country will from this benefit from
this comprehensive environmental gardening manual.

Table of Contents on next page

TABLE OF CONTENTS

- -

Please send ____ copies of "Designing and Maintaining Your Edible Landscape - Naturally" to:

Name_____

Street Address_____

City _____ State _____ Zip _____

Each book is **$20**, plus **$4** shipping and handling. California residents add $1.45 sales tax.

Please make your check payable to "The Edible Landscape Book Project."

Mail to: The Edible Landscape Book Project, PO Box 1841, Santa Rosa, CA 95402.

Gray Water Use in the Landscape
how to help your landscape flourish with recycled water.

25 PAGES • 11 ILLUSTRATIONS • APPENDIX OF SAFE SOAPS & DETERGENTS

by Robert Kourik

Gary Gray Water Sez,
Gray Water is Grayt™

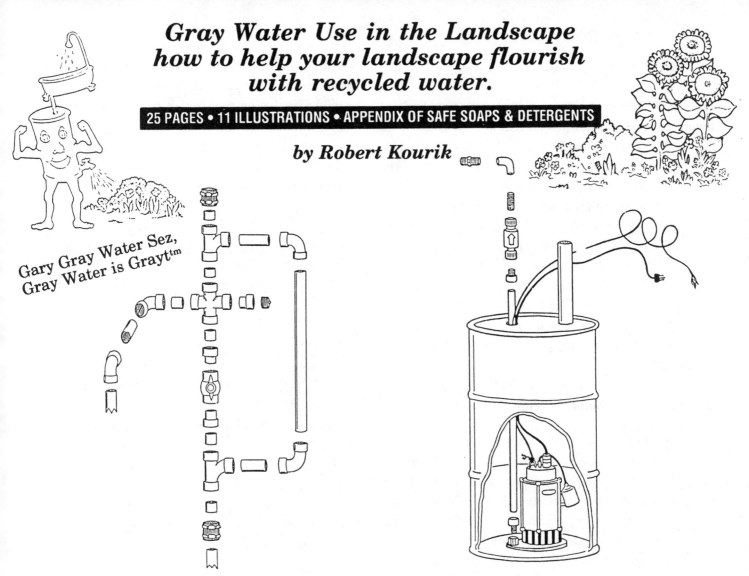

Plumbing options and parts
are fully described.

A pumped gray water system.
(The drum is a temporary buffer or
surge tank, not for long-term storage.)

A "grayt" help to gardeners everywhere.

*Gray water makes plants grow better than with city or well water
because it comes with fertilizer already added!*

Also, a "grayt" help to drought-plagued gardens.

Please send _____ copies of "Gray Water Use in the Landscape" to:

Name _____

Street Address _____

City _____ State _____ Zip _____

The total cost is $6, which includes 1st Class mail shipping, handling and,
in California, sales tax.

Please make your check payable to "Edible Publications."

Mail to: Edible Publications, PO Box 1841, Santa Rosa, CA 95402.

Survey

With each publication, I survey my readers to find out their interests, what they want to know about and what information they can't find. Feel free to write suggestions on topics for future handbooks on the bottom of the page. If you return a copy of this survey, you'll be the first to know when I publish something on the topic you're interested in. If you've checked off the box for the newsletter, and I end up publishing it, you'll get a prepublication discount notice when the newsletter goes to the printer. Due to the demands of self-publishing, I can't respond to each note or survey individually. I will file it in the proper place, and you'll be the first to know when your favorite topic is "hot off the press."

Bob's Honest-to-Goodness Newsletter

Most gardening magazines are now driven by their advertising departments. The ad guys (and women) actually review article concepts to screen for those which might threaten the advertisers. Consequently, I've had a number of article ideas turned down because "we're losing too much money and we don't want to threaten our ad revenues." So how about a sporadically produced newsletter which allows me to give you the straight gardening poop which no other magazine will? Opinionated straight talk, no b.s.

Also, I'm interested in a wide range of oddball and eclectic topics which can't stand on their own as independent newsletters. So I envision Bob's Honest-to-Goodness Newsletter as a freewheeling discourse on many topics: lavender as a plant and cooking ingredient; constructed archaeological garden sites (fake archaeological digs based on your own mythology); lithium and its impact on manic-depressives; the myths of the ecological movement (dogmatic rhetoric is not the domain only of the political right or the anti-environmental corporations); the dark side of nature (wildness is not just the pristine cuteness and scenery found in glossy environmentalist magazine articles) and tirades against those whose environmental perspectives are ethnocentrically, narcissistically and anthropo-morphically based. It's about time the ecological move-ment embraces true diversity of opinion and a bit of fuzzy logic.

Because there will be no advertisers to suck up to, the newsletter won't be cheap. But it'll certainly be fun and provocative.

Dear Robert,

☐ Yes, I'm interested in your newsletter. Keep me posted. Name: _____

Address: _____

Also, I'd like to see a publication on the following topic(s):

_____

_____

_____

_____

Mail to: Robert Kourik, P.O. Box 1841, Santa Rosa, CA 95402.

KOURIK DRIP IRRIGATION SYSTEM

This is an innovative state-of-the-art drip irrigation system that meets the needs of home gardeners and landscapers. It uses the highest quality components to ensure excellent performance. The Kourik System is also easy to install, modify, and maintain. It comes with a pre-assembled Master Unit that contains all components essential to a complete drip system, and unique tubing with pre-installed emitters. Use of the Kourik System provides substantial water savings and enhanced growth.

The Kourik System works well for:
♦ Vegetables or flowers
♦ Rows or beds
♦ Trees, shrubs, or hedgerows
♦ Flat areas or slopes
♦ New lawns
♦ Long, narrow areas that cannot effectively be served with sprinklers.

ADVANTAGES OF THE KOURIK SYSTEM

1. **Easy installation:** Most drip systems require a laborious and messy installation with clamps, emitters, and a mass of spaghetti tubing—aptly named, because of the tangled mess it can become. Once installed, there are exposed parts which can, and do, snag or break. The tubing or drip lines for the Kourik System come with the emitters already neatly embedded inside. There is no spaghetti tubing, and no tools are required other than a knife or heavy scissors. All you need to do is screw the Master Unit onto your faucet, connect it with a length of hose to your drip lines, lay them out, turn on your tap, and enjoy the results.

2. **Unique Emitter Tubing:** Tubing with pre-installed emitters is an important advance in drip irrigation because it saves you time, effort, and aggravation, and it irrigates better. It was developed based on the latest horticultural knowledge showing that plants grow best when the soil around them is evenly watered, not when each plant is treated as an island. The Kourik System's emitter tubing spaces emitters at 12 inch intervals, thus offering an intelligent improvement over other systems which direct you to install one or more emitters at each plant. Its specially designed emitters are virtually clog-proof.

The Kourik Drip Irrigation System can save you time, water, reduce weeds and increase yields as much as 84%.

They are also pressure compensated, to deliver equal water to all plants over distances up to 300 feet or on any degree of slope. As a final bonus the tubing is brown for camouflage on surface applications.

3. **A Pre-Assembled Master Unit:** The Master Unit is a unique feature of our System. It contains the components essential to give you all the advantages a drip system can provide: a twin-tap connector, so you can screw the system onto one tap and still have another free for other purposes; a backflow preventer to keep contaminents out of your household water; a filter to keep emitters from clogging; and a pressure regulator for proper flow of water from your tap into the drip system. All components used in the Master Unit are top-of-the-line. Because they are pre-assembled in one compact unit, you need only twist the unit onto your hose tap.

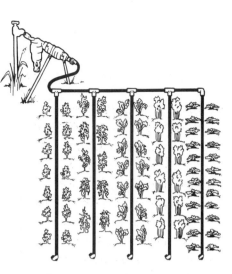

Two Ways to Purchase Your System

The Kourik Drip Irrigation System can be purchased as a kit or you can buy the components separately and design your own system. Either way, the Kourik System is easy to modify and expand so you can meet the changing requirements of your garden and landscape. All systems come with illustrated, step-by-step instructions.

The Kourik Drip Kit

These are for gardeners who prefer kits, or for gardens laid out in a conventional row or bed design. Kourik Kits come with a Master unit, 25 feet of solid feeder line (solid hose to get water from your faucet to your garden), emitter tubing in lengths of 100 feet, 200 feet, or 300 feet, and a set of easy-to-use fittings. These systems can be added to and modified at any time with components available separately.

100' System (fittings for up to 5 rows) #940 $109.00

200' System (fittings for up to 10 rows) #941 $149.00

300" System (fittings for up to 15 rows) #942 $189.00

Components

For those who want to design a custom system or modify one of our kits, we carry a complete line of components.

100' Emitter Tubing #943 $39.50

1000' Emitter Tubing #952 $385.00

Master Unit #944 $65.00

Tees #945 $.65

Elbows #946 $.55

Straight Coupling #947 $.40

End Closures #948 $.20

J-Stakes #949 $.10

Solid Feeder Line, 25' #950 $ 6.25

Threaded Tees #951 $2.05

Add-It Fertilizer Injector

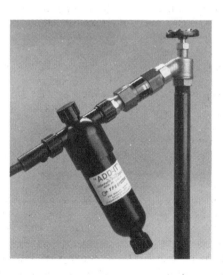

The Add-It Fertilizer Injector attaches easily to any drip system and simplifies fertilizing your landscape and garden. Just fill the cannister with water soluble fertilizer and turn on the water; the Add-It automatically mixes fertilizer into the water stream and distributes it evenly throughout the drip system. Can be used with overhead sprinkler systems. To use with the Kourik System, please match the Add-It model with system size.

Add-It pint (100' — 400' system) #962 $19.95

Add-It quart (200' — 800' system) #963 $37.95

Add-It 1/2 gallon (200' — 800' system) #964 $55.95

Add-It gallon (500' — 1500' system) #965 $109.95

Nelson Electronic Water Timer

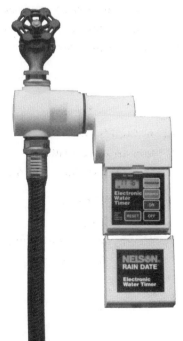

The Nelson Electronic Water Timer automates your lawn and garden watering, freeing you from daily watering chores, and allowing you to take vacations knowing your garden will receive the water it needs. It is ideal for sprinklers and drip irrigation systems. The Nelson Water Timer is easy to program, and may be set to provide up to three separate watering cycles per day, seven days a week. It also has a manual override button for occasions when you wish to water outside the programmed schedule. The water computer will run for a full season on four AA batteries, and should you ever need assistance, comes with an #800 toll-free help-line.

#913 $59.00

Index